SMARTER GOVERNMENT

How to Govern for Results in the Information Age

MARTIN O'MALLEY

Esri Press
REDLANDS | CALIFORNIA

Esri Press, 380 New York Street, Redlands, California 92373-8100
Copyright © 2019 Esri
All rights reserved.

Printed in the United States of America
23 22 21 20 19   2 3 4 5 6 7 8 9 10

Library of Congress Cataloging-in-Publication Data
O'Malley, Martin, 1963-
    Smarter government : how to govern for results in the information age
    LCCN 2018052375 (print)—LCCN 2019006940 (ebook)—ISBN 9781589485259 (electronic)—ISBN 9781589485242 (pbk. : alk.
    paper) 1. Internet in public administration--Maryland. 2. Geographic information systems--Maryland.

    LCC JK3849.A8 (ebook)—LCC JK3849.A8 O53 2019 (print)—DDC   352.3/80285--dc23

For the young people of the United States of America
and the generous and compassionate nation they carry in their hearts.

Contents

Foreword

Over the last fifteen years, I have directed the Innovations in American Government Awards and taught at the Harvard Kennedy School in Cambridge, Massachusetts. In the early years of my work there, students viewed a job in the federal government as the highest calling. Today, many more of those graduating students see local governance and local problems as the place to apply their talents. And these highly trained students need both what they learn in our classrooms, as well as inspiration from leaders who have made a difference. The author of this book, Martin O'Malley, is such a leader, and has been one consistently over the last decade and a half. During that time, he has unfailingly advocated for leadership and management reforms that would improve the quality of life for residents of Baltimore and Maryland.

In 2004, I presented Harvard's top award for innovation in government to then-mayor O'Malley for his work with CitiStat. His words at that time frame his subsequent accomplishments and many of the themes in this book. He explained how he sought to introduce tools and procedures that, in his words, would "transform the way our city government works," moving past political patronage to a results-based system. Even then, his attention was on mapping the quality of services to ensure equity and uniformity in services. His goal was to marry a new operational corporate culture with his progressive approach. In a thread from that presentation to this book, O'Malley emphasized that performance management was also a way to identify high-quality public employees.

When I visited CitiStat in the early 2000s, I saw the now-famous interaction between the agency directors and the data-driven team that ran the process. But even more, I witnessed an engaged mayor who measured and demanded results—just the sort of inspiring conduct that has made O'Malley, both as mayor and as governor, an example for students who are training for careers in public service and for current leaders.

In the years I have been working with governors, county executives, and mayors, I have seen a wide range of talents. There are leaders who charismatically rally the city, those who work well with communities, those who can manage well, and those who lead well. But what Martin O'Malley brings in person and in this book is the full complement of those skills. In addition to addressing specific issues of streets, air quality, education, and crime, he shows us how to combine governing, leading, and managing. These lessons will be broadly important to our cities and states today where so many residents depend on high-quality, honest governance. At a time of rising cynicism, this book provides an upbeat optimism about what those dedicated to improving the quality of life can produce by combining high-quality leadership with an unrelenting commitment to operational details and excellence.

—Stephen Goldsmith

Stephen Goldsmith is the Daniel Paul Professor of the Practice of Government and the Director of the Innovations in American Government Program at the Harvard Kennedy School. He served as the 46th mayor of Indianapolis and also as the deputy mayor of New York City for Operations.

Preface

A new way of governing is emerging. It is rising from cities, counties, and towns. It is a quiet but rapidly developing evolution—an evolution of enormous positive potential for thinking, caring, and rational self-governance.

Unlike the hierarchies of command and control that characterized the old way of governing, this new way is collaborative, entrepreneurial, interactive, and relentlessly performance-measured. And it is enabled, for the first time in human history, by powerful new technologies of geospatial intelligence and data sharing.

The combination of geographic information systems (GIS) and the Internet of Things (IoT) has given us the ability to model complex systems, measure performance, and see and measure what works on a scale—and with a timeliness—never before possible.

Whether the goal is improving public education, reducing violent crime, or restoring the health of our natural environment, GIS provides a powerful platform for progress. GIS is making government smarter.

But technology by itself is not enough.

Effective leadership is essential.

Effective leaders in the Information Age create common platforms for collaborative action. They focus the problem-solving dialogue on the emerging reality displayed on dynamic maps. They develop routines for convening leaders around this platform to measure effectiveness, lift up successful techniques, and understand better ways of getting things done. They pull data from the shadows of traditionally isolated bureaucracies to create a vivid and dynamic picture of the whole. They use geospatial intelligence to drive innovation and accountability into the center of the collaborative enterprise we call "governing."

First as mayor of Baltimore and then as governor of Maryland, I experienced firsthand the power of GIS and performance management. We adopted and adapted the "stat" approach—first pioneered by Jack Maple with CompStat for the New York Police Department—into CitiStat for Baltimore, and later into StateStat for Maryland. With data, the map, and this new method, we tackled huge challenges and made nation-leading progress in the face of big, complex problems.

In this book, I share in common-sense language and compelling personal stories how any modern leader can use these same principles and methods to achieve goals, lift up its high performers, drive effective collaborations, and transform often-moribund organizations into higher-performing teams. This book is not so much a user's manual for Stat or GIS, but rather a practitioner's guide for collaborative leadership in the Information Age.

This book is about the data, the map, and the method for achieving dramatic public-sector progress. It is about making complex problems visible and understandable for everyone who has a stake in seeing better outcomes and results. This book makes a new and better way of governing simple, demonstrable, and understandable for every citizen who believes their government should work to deliver better results. But really it is about serving people well.

This new, smarter way of governing is not about left or right; it's about doing the things that work to move us forward as individuals, as communities, and as a people. At the end of the day, it's all about making better choices for better results—results that make a positive difference in the lives of every individual, in the lives of our kids, and in the life of the common good we share.

We do this by setting clear goals, measuring progress, and getting things done.
We start and don't stop.
We lift up the leaders.
And we lead.

—Martin O'Malley
  May 2019

Introduction

At the age of thirty-six, I was elected one of the youngest mayors in the history of the City of Baltimore. It was not a path that anyone might have predicted for me a year earlier. To be clear, I did not run because my city was doing well. There were no cheering throngs urging me to run. I ran because my city was bleeding to death and someone had to stop the bleeding.

By 1999, Baltimore had become the most violent, addicted, and abandoned city in America. The attitude in most quarters of our city at that time was one of resignation—a sinking sense that our problems were bigger than our capacity to solve them. This despair was not restricted to downtown leadership or the philanthropic boards of mostly white people. The lack of belief was particularly acute across the poorest neighborhoods of our city where violent crime claimed the lives of more and more young black men per capita than any city in America.

Against this reality—and with only eighty-eight days until the Democratic Primary—I decided to run for mayor.

For a Better Baltimore

Baltimore then, as now, was a majority African American city, and I was a white candidate. My two primary opponents in the race for mayor were both African American men—both were older and far better known than I was. One was the city council president directly elected from the hard-hit west side of town. My other opponent was an outspoken council member who had run for city council president; he was from the similarly hard-hit east side of town.

I knew them both very well. We served together as elected members of the Baltimore City Council. But I sensed that neither of them could do what needed to be done to turn around Baltimore's violent crime problem. And I sensed I could. After serving for eight years on the city council, I felt ready to leave public service and throw myself into a higher-level practice of law. But as a former prosecutor, I had also become one of the leading voices on the city council for improved public safety and criminal justice reform.

My gut told me the public had finally grown sick and tired of being sick and tired. And we were looking for new leadership.

On the council, I had watched with awe and envy as New York City reduced violent crime to record lows, even as national television shows were made about Baltimore's seemingly intractable crime problem. We learned how the New York City Police Department was using a new performance-measured approach to policing which they dubbed CompStat. Putting dots on a map to show where crime was happening in real time, regardless of race. Using new mapping technology to better deploy detectives and patrol officers in more timely and effective ways. Bringing commanders together in collaborative circles with a regular cadence of accountability around the emerging truth of where and when crime was happening. Taking better actions to save more lives.

My campaign—our campaign—was not about excuses or scapegoating. It was about real solutions to the real big problems of violent crime and drug addiction. In a programmatic sense, it was about bringing CompStat and a new way of policing to Baltimore. But in a spiritual sense, it was about something deeper.

Our campaign was about justice and injustice and the freedom to choose. The justice of feeling safe in one's own home and neighborhood—regardless of race, class, or place. The injustice of allowing violent crime and 24/7 open-air drug markets to become the new normal across the poorest neighborhoods of our city. The injustice of a citywide apathy that had so many of us—regardless of race—shrugging our shoulders and behaving as if there were nothing that could be done about drug addiction and violent crime.

And at its core, our campaign was about the freedom to choose a different, better future.

A New Way of Governing

On Election Day, we won every council district in the city, defeating my two primary opponents in their own districts. In the General Election, we won 91 percent of the vote. When I was sworn in on December 7, 1999, we quickly brought CompStat to Baltimore policing. And as we put Baltimore on a path for the biggest ten-year crime reduction of any major city in America, we also set out to bring this new way of governing to the whole of city government.

We dubbed this new way of governing CitiStat. It earned the Innovations in American Government Award from the Harvard Kennedy School in 2004. If you were to search on *CitiStat* today, you would find it popping up with thousands of entries across the country and indeed around the world. In fact, almost every major city today in America now operates by way of a 311 call center for receiving citizen complaints on the front end and some degree of performance management on the back end.

Whether it is fighting crime, filling in potholes, or ensuring the trash gets picked up on time, these and a thousand other tasks are the day-to-day operations that make up the work of any city government. Now, thanks to new technologies—primarily, geographic information systems (GIS)

Inauguration Day, December 7, 1999—walking to the podium across the cobblestones in front of City Hall with Katie, Grace (8), Tara (7), and William (2) in Katie's arms.

and the Internet of Things (IoT)—these issues can all be mapped, managed, and measured with greater speed and accuracy than ever before. City services can be delivered with greater openness and transparency than ever before. Our governments can operate with greater efficiency and effectiveness than ever before.

And it's not just for cities.

The Power of Information Shared by All

Say the words "separate silos of information" in government circles anywhere in the English-speaking world and see if it doesn't make heads shake and eyes roll. It is a time-worn cliché. Whenever big problems need to be addressed, we quickly start to bemoan the existence of "separate silos of information."

When it comes to solving dynamic and complex problems, the separateness of information can make coordination and collaboration nearly impossible. In an emergency, the separateness of operational and situational information can have fatal consequences. And unfortunately, this dysfunctional separateness remains a pervasive fact of life that many leaders in and out of government face daily.

Many of us will long remember the attacks of 9/11. We remember how the New York City Police Department—aware the World Trade Center Twin Towers were about to fall—was ordering its personnel to evacuate the area, even as the New York City Fire Department was ordering courageous firefighters—who never got the message to evacuate—to climb up the stairs of the burning buildings. The after-action reports called it "a lack of interoperable communications." A tragic failure to communicate.

The first time I met Esri co-founder and president Jack Dangermond was in Annapolis in 2007. I was by that time a newly elected governor. Jack had kindly asked for the meeting. And knowing of Jack's pioneering work in the field of GIS technology, I asked a few of my senior staff to be there, as well.

"You've already figured out something as mayor that most elected officials have yet to figure out," he said.

Incredulous but longing for affirmation, I asked, "What is that?"

Then—with an assortment of coffee cups and water bottles—Jack took us through a lesson in governing in the Information Age that I have never forgotten.

"Imagine," he said, "that each of these coffee cups or water bottles is a different department of your government..."

He then proceeded to commandeer our assorted drink containers—naming each one as he took them from our grasp. "This one is the Police Department... this one is the Fire Department... this one is the Health Department... this one the Housing Department..." He assembled them into a haphazard grouping on the conference table in front of us as my chief of staff quickly grabbed one last hit of caffeine before relinquishing his coffee mug.

"Each of these departments has their own separate organization, with their own command structure. And, importantly, their own separate silos of information—their own separate data and information about what they do, how they do it, and where they do it..."

(The unorganized grouping of separate containers that didn't speak to each other was an executive branch metaphor all of us understood painfully well.)

"Now," he proceeded, "you could spend millions of dollars, and years and years of time, hiring information technology contractors to try to connect up and down each of these separate silos of information. You could try to create the IT fix to translate different data collection methods from different formats. You could pay IT consultants to make sure the right people up and down these different departments know all the time what their colleagues in other departments are doing and how those actions might impact their own department's mission and their work.

"You could try that," he said, "but you'll waste a lot of money, you'll waste a lot of time, and you'll never get it done. Or... give me your legal pad..." he said.

"Or, you could simply use one map and insist that the databases—of each of these separate silos of information—land on the same map."

And with that, he calmly placed the bottom of each cup and bottle on top of my legal pad turned GIS map.

"Now, a picture emerges on the map. Everyone can see what everyone else is doing and where they are doing it. As long as there is an address for an activity, the map integrates all the different actions from different agencies. It becomes a single picture for understanding and seeing many dynamic pieces. The departments can keep whatever format and data collection methods they like. They just need to be open and land the bases of their data on the map.

"The map creates the picture. And it is a picture everyone can see and understand."

The map can also tell us about things that are far more stationary and static than the deployment of emergency personnel. The map can tell us where the roads and highways are. Where the people live. Where the water infrastructure is. What neighborhoods will be underwater with the volume of rain that falls in a 100-year flood. Or which streets will be inundated when a Category 4 hurricane slams into your city.

Belief Space, GIS, and the IoT

What Jack showed us with cups and bottles on a legal pad was a "common operating platform." For members of the Uber generation, the wonder is that all of government doesn't operate by way of common platforms just like ride-share and bike-share companies do. But the truth is, this new technology is still emerging in the operations of most of our governments. Fast evolving, but still new. Some leaders have figured it out more quickly than others.

The capacity that a common operating platform provides for managing dynamic and sometimes fast-moving problems with real-time data is a big innovation in government. In fact, these new technologies—GIS and the IoT—and the ability these technologies give us to model belief space—that is to say, to model the changing dynamics of our built and natural environments—are ushering in a whole new way of governing in the Information Age.

People born after the early 1990s might think these technologies have been here forever. So many of us who use Uber act as if we could always call for a car from anywhere and watch the little vehicle icon weave its way across the real-time map on our phone to pick us up. But this technology is all still very new for governments of, by, and for the people.

More than fifty years ago, Robert F. Kennedy famously said, "It is from numberless diverse acts of courage and belief that human history is shaped." That was true when he said it, but it is not quite so true today. In the Information Age, those diverse acts of courage and belief are no longer numberless. They can all be numbered. They can all be measured. They can all be mapped. In fact, we can measure the ripples in real time to figure out how many it will take to make a wave of change.

Let me show you what I mean.

This is a screenshot from my iPad, of my weekly commute.

My morning rush-hour drive from my home in North Baltimore to my class at Georgetown University on the far side of Washington, DC. This is one everyday example of our newfound ability to model our natural and built environments and the human dynamic that plays out over it in real time. In this case, the challenge is traffic avoidance for on-time arrival. But our ability to model belief space and to measure the movement of dynamic systems—in real time, with probabilistic certainty—are big game changers in the effectiveness of public administration.

Every week during one recent fall semester, I had to travel during rush hour from North Baltimore across two big metro areas to teach at Georgetown University on the far side of Washington, DC. If you were to tell my neighbors where I was headed on those mornings, they would likely say, "Oh, you can't get there from here at that time of day!"

So how do I get from point A to point B? I do what we all do now. Without even thinking about it, I plug in the address into my iPad, and a pleasant-sounding female voice tells me exactly which way to go and how to get there quickest. I call her, "Mary," and Mary is never mistaken about the road or the destination.

Sometimes, conditions change on the route ahead—road closures, accidents, the speed with which other people are traveling along the road ahead of me. But today, all those things get uplinked immediately through the IoT. They come back to my map in real time, so Mary can tell me, "A quicker route is available."

In this example, the dynamic belief space I had to navigate was represented on a two-dimensional (2D) highway map. Different datasets were integrated together on that map to model a way forward with a probabilistic certainty of outcome: arrival at a destination within a certain time. The dynamics of the natural and built environment were brought together into an operating picture from multiple datasets.

But what if the destination were eradicating childhood hunger, improving educational outcomes across a state, or improving the health of our people, our land, our water, our air? These same new technologies are now being brought to bear on all these critical pursuits.

Geographic information systems. The Internet of Things. Our ability to model belief space with probabilistic certainty.

These are the tools that unlock a host of new possibilities for collaborative leaders in the Information Age.

These are the new technologies of a new way of governing.

1

A New Way of Governing

There is a balance—a symmetry, if you will—between how well we govern ourselves and how much we are capable of trusting one another. In a democracy, there is an inseparable relationship between the two. They can reinforce each other, build each other up, or drag each other down.

A New Formula for Effective Governance

Sometimes positive change is hard to perceive when it is happening close to you. You won't see it proclaimed across the 24/7 "breaking news" tickers, but a new and better way of governing is emerging across our country.

If a lack of trust is the greatest political challenge we face as a self-governing people, perhaps this new way of governing holds the promise of a better way forward.

It is evolving in some places faster than in others, but the movement is undeniable. It is rising from cities and counties to states. And hopefully one day soon, it will make the leap from state governments to our national government. It is a more effective kind of public administration—a new, information-enabled way of governing—that many mayors and county executives across America are bringing forward to better deliver results. These results are building up mutual trust among citizens in our cities, towns, and metro areas—places where civic trust is actually on the rise.

Today, most Americans feel a lot better about how their cities, towns, and local counties are governed than they did just fifteen years ago. And it is the direct result of the emergence of this new and better way of governing that is based on:

- Performance management and data-driven decision-making
- GIS technology
- Customer service technology, such as a single phone number for citizens to call for city services (311)
- Collaborative circles of caring people who are making decisions based on the latest emerging truth about what works, rather than on the old habits of "the way we have always done it"
- Openness and transparency
- Getting things done by bringing people together regularly to think, question, and act in more effective and collaborative ways

The formula for effective democratic governance in the Information Age requires a radical commitment to openness and transparency that is demonstrated every day through actions, not words. It lifts up effective collaborations over rote obedience to command and control. It provides real-time feedback loops enabled by modern technologies and the internet. It creates a rapid cadence of accountability—a cadence laid down by the discipline of short, regular meetings of stakeholders who are focused on the latest emerging truth. It is the ability to model "belief space" about our physical world with probabilistic certainty. It is an eco-systemic approach to understanding an array of actions and interactions—how they impact our natural environment, how they shape our built environment, and how they advance the common good we share.

Whether we are talking about reducing crime or reducing air and water pollution, we are talking about systems. And although these systems often are interconnected across city, county, and state borders, they are not infinite systems; they are closed systems. They can be mapped and measured. The causes and effects of positive and negative feedback loops can be modeled, anticipated, and changed by the actions we take.

The Changing Nature of Authority in the Information Age

There was a time, not so very long ago—before cell phones and the internet—when leaders knew things days, weeks, and even months before everyone else got the news or could figure it out on their own. In the old days, things got done according to the decree of "because I said so." It was all about authority: the authority to know and the authority to give orders; the authority to force and to enforce.

In those days, the place of positional advantage for the leader was to stay high atop the hierarchy of command and control. From this position, leaders had a distinct advantage. Information from multiple sources flowed up to the leader, who was able to put it all together before anyone else could. As information flowed up, orders flowed down through various chains of command and control.

The Information Age is rapidly changing the relationship between leaders, people, and information. Leaders can no longer control information or the timing of its release.

What Is GIS?

Maps and data layers, both 2D and 3D, underpin GIS, a technology that organizes information into all types of layers that can be visualized, analyzed, and combined to help us understand almost everything about our world. For example, GIS incorporates all kinds of data layers about initiatives in our government and their impact on citizens, enabling everyone to better understand our situations, our scenarios, and our decisions.

Because all GIS layers can be overlaid and integrated using maps and geographic analysis, modern GIS systems living on the web provide everyone with a universal integration engine to better understand and manage our operations and activities.

Today, hundreds of thousands of organizations in virtually every field of human endeavor use GIS to make maps that help us understand, communicate, perform analysis, share information, and solve complex problems. The use of GIS is literally changing the way the world—and our governments—work.

GIS integrates data about everything—rooms in a building, parcels of land, infrastructure, neighborhoods, local communities, regions, states, nations, our planet, and beyond—to other planets. The GIS nervous system provides a framework for advancing scientific understanding and for integrating and analyzing all types of spatial knowledge. Because all layers share location as a common key, any data theme can be overlaid and analyzed in relation to all other layers that share the same geographic space.

The idea of georeferencing shared data is a powerful notion. Suddenly, it's not just your own layers or the layers of your colleagues that are available to you—it's everything that anybody has ever published and shared about any geographic area. This capability is what makes GIS such an interesting and useful technology; you can integrate any of these different datasets from a range of data creators into your own operational views of the world, overlay them, and perform spatial analysis to derive deeper insights and understanding.

Learn more at GIS.com.

Thanks to the internet, social media, and cell phones, We the People now know as much as our leaders, and we usually know it *before* our leaders.

This is a perilous time for leaders who are information control freaks.

Of course, we still need command and control in functional governments. We still need bureaucracies and hierarchies. Rank and lines of authority are still important; in some situations, such as military conflicts or emergency relief efforts, they are essential. But even in emergency situations today, people usually know what's going on at the same time or even before many of their leaders do.

When titles of authority no longer guarantee the earliest or best information, authority no longer guarantees legitimacy. Simply being in charge is no longer enough to maintain public trust. For authority to be legitimate, authority must be effective. And it must be demonstrably so.

The only place of positional advantage for a modern leader today—and the only place from which a leader can be truly effective and legitimate—is not from high atop a pyramid of command and control, but rather from the center of the latest emerging truth—the ground truth; the truth on the map. The map which can be seen by all.

Today, legitimacy derives less and less from title and more and more from doing what works to achieve demonstrably better results—results that people can see as clearly as their leader can. To be trusted, leaders must know and leaders must be able to show.

Is crime being driven down or is it increasing? Are more jobs being created than lost? Is student achievement improving or is it declining? Do you know? And are you doing something about it? Effective leaders can answer these questions only if they are close to the emerging truth. If they are consistently holding the center of the collaborative circle—people whose perspectives and expertise allow for the best possible actions to be taken, in the moment, to affect and change the dynamic conditions on the ground.

In the Information Age, effectiveness also requires a leadership commitment to openness and transparency—a commitment and practice that until this day might have seemed radical, or even

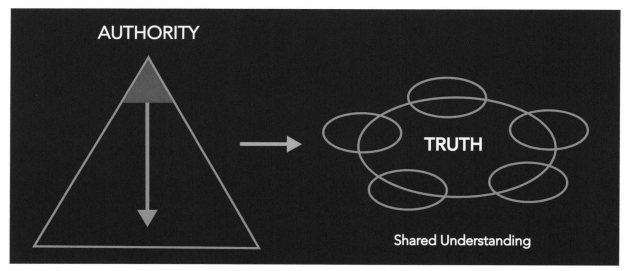

In the Information Age, people know things at the same time as their leaders. Therefore, the place of positional advantage is no longer high atop a pyramid of command and control where information can be tightly held, but rather, at the center of a collaborative circle, focusing the attention of responsible leaders and stakeholders on the latest emerging truth.

politically naive and reckless. Setting public goals with public deadlines, measuring performance in ways that are shared by all—these are the hallmarks of a new way of governing in the Information Age.

The Power of Maps That All Can See

In this new way of governing in the Information Age, the GIS map plays a central, organizing role. It is not just a nice picture or another layer. The map—the geographic information map—becomes not only the integrator of once-separate silos of data, but also the field of action on which effective collaborations and winning plays can be run. In well-led governments across our country, The Science of Where® is empowering the art of how in public administration.

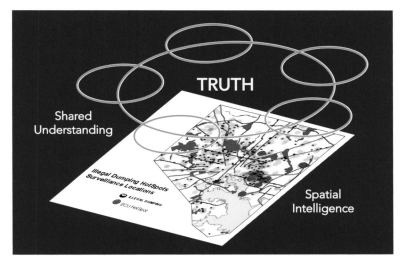

In the old days, things got done according to the rule of "because I said to." But today, title and authority are not enough. We must be able to show one another what works, where, and why. It's all about shared understanding. Spatial intelligence puts the latest emerging truth "on the map" for all to see.

Let me explain.

On an American football field, there are eleven players on offense and eleven players on defense. The

What Are Open Data Portals?

Open data portals are central locations where governments can store data and make it easily accessible to the public. These portals increase government transparency and accountability by providing citizens with unprecedented levels of access to their government.

When I was governor of Maryland, my StateStat team managed Maryland's Open Data Portal—an online database of more than five hundred searchable, machine-readable datasets uploaded by various agencies across the state. StateStat used the data in the portal to track progress toward our sixteen strategic goals. As one of the few states that link progress directly to open data, Maryland led the nation in government transparency and accountability.

In 2015, I signed SB 644 into law. The law requires Maryland's state agencies to publish open data and to establish a Council on Open Data to drive progress forward. The Center for Data Innovation recognized our efforts, and named Maryland one of the top states in the nation for open data.

goal is to move the ball all the way down the field to score touchdowns and field goals. Each player can run, block, pass, or catch. But to move the ball down the field, those activities must be done in coordination with one another. The activities must be synchronized. There are brief huddles in which plays are called. Different plays must be run against a dynamic and changing defense. Every part of the field has a geographic coordinate—yard lines, sidelines, hash marks, and end zones. The movement of the ball is measured over the space-time continuum. And for the entire time, the clock is ticking off the minutes and seconds that remain in the game.

In the Information Age, GIS gives leaders the ability to turn their entire city, state, or country into a highly visible, and accurately measured field of play. And the stakes are far greater than scoring points in a football game.

Whether the goal is reducing violent crime, eradicating childhood hunger, improving educational outcomes, or improving the health of an entire ecosystem, GIS provides the ability to unify separate efforts, actions, and data into a common operating picture for all; an operating picture that tells us whether the plays we are running are actually moving the ball down the field. And today, most of this can now be seen in real time by everyone, both inside and outside government.

Call it what you will—spatial intelligence, GIS, dynamic maps, smart maps. This type of information technology gives us the ability to see, measure, and manage complex systems over the space-time continuum in real time. And this is a huge, new development for effective public administration—for our ability to better govern ourselves.

The Four Tenets of Performance Management

The four tenets of any performance management regimen or "stat" process—such as CompStat, CitiStat, StateStat, and others detailed in this book—are:

1. Timely, accurate information shared by all
2. Rapid deployment of resources
3. Effective tactics and strategies
4. Relentless follow-up and assessment

These four tenets are not end points. They are the beginning and the way. They are the constant framework of an ongoing search for better and more effective ways of collaborating and getting things done. The pursuit requires constant thought, intellectual curiosity, and leadership that is open to bringing forward the right questions—wherever they might lead.

There will never be a point where the information is as timely and accurate as it can ever be; there will never come a time when the organization arrives at the perfect combination of tactics and strategies that never need to be changed again. Improvements, pivots, and adaptations never stop.

So, don't let the impossibility of "perfect" be the excuse for not starting.

No organization or leader ever started with timely, accurate information across the whole range of endeavors for which they are responsible. But good leaders and performance-managed organizations can use these four tenets to *improve* the timeliness, *improve* the accuracy, and *improve* the openness of their information every day. Bottom line: you start where you start and you get better.

Rapid deployment of resources—whether to intervene in a health challenge or sanitation challenge or education challenge—is never as rapid in the earliest days of confronting a problem as it can be made over time.

Whether tactics and strategies are effective is a question that must be asked every day in light of changing realities, changing actions, and the changing external dynamics of almost every human problem.

Effective public administration remains a longitudinal experiment in our time, just as it was in George Washington's time. As such, it demands relentless questioning, relentless doing, and relentless follow-up in the face of changing circumstances. Today, we have far better maps and technologies than George Washington might ever have dreamed, but the longitudinal experiment—the pursuit of better—is not an activity that stops when inauguration or transition is over.

Nor is it like cramming for a test. The search for better is a daily pursuit; it is a daily organizational discipline.

Creating a Cadence of Accountability

There are seasons and cycles to everything in nature. And so it is with democratic governments. Many of these traditional cycles are well-known to us as citizens. There are election cycles. There are legislative cycles. There is an annual budget cycle. And in larger cities, there are "monthly closeouts"—smaller budget cycles printed in little books that show how money has been spent on every line of an annual budget by month.

At the local level, there are schedules for trash collection, leaf collection, bulk trash collection, and other services. At the state level, driver's licenses, car registrations, and hunting licenses must be renewed from on set annual cycles. But, in between revenues required and services delivered, there is the gray zone of "bureaucracy"—a land that seemed in the past to defy cycles and measures.

Marc Morial, the former mayor of New Orleans, said to me as a new mayor (only half-jokingly), "Kid, if

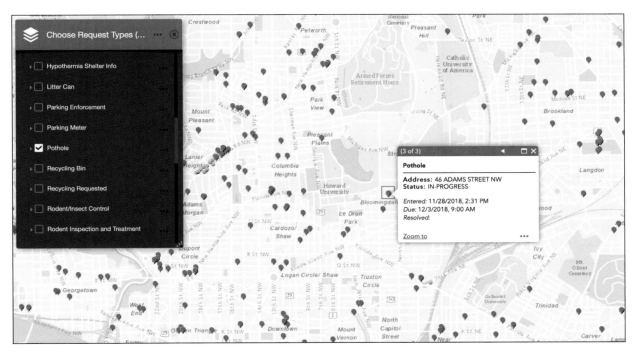

A new way of governing has quietly taken root over the past decade in local and municipal governments. Cities across America have adopted geographic information systems and a single phone number (311) for city services. Here, the District of Columbia's 311 Map allows the public to see 311 customer service requests (in this example, potholes) in the last 30 days by location and status.

you ever want to hide something in city government, make sure you put it in the city charter or the city budget—because nobody ever reads either one."

Our governments became very adept at measuring annual revenue inputs, but not so good at measuring weekly operational outputs—like, how much trash did we pick up? How many crimes did we solve? Or, how many abandoned vehicles did we tow? Let alone whether we were delivering those services more efficiently this week than last week. Was absenteeism this week up or down compared to last week? Was the use of overtime up or down? Who knows? Who can tell? "We could find out, but we would have to pull all our people off their other jobs, and it will take weeks . . ."

Not anymore. The technologies of the Information Age are stripping away those excuses. Armed with the information and the ability to map it in real time, we can now turn information into operational wisdom, operational understanding, operational foresight, and operational change. Doing this requires a *cadence of accountability*—regular, predictable convenings of responsible players around the maps and graphs of the latest emerging truth about conditions on the ground and what we are doing to change those conditions.

These regular, short, focused operational convenings are not meetings simply to have meetings. Meetings are not actions. But unless we meet—with openness and honesty about the evidence before us—we cannot run the plays that make progress. These "stat meetings" are valuable, priceless, single hours to focus the collective intelligence of the enterprise on solving problems, adjusting and improving, and driving leading actions to the goal.

All people work against deadlines. But only the leader can lay down this cadence of accountability.

If output is measured annually, people will make incremental annual progress. But if progress is measured every two weeks, people will make incremental progress every two weeks. Two percent improvement annually is mediocre. Two percent improvement every two weeks can be nation-leading progress.

It is up to the leader to lay down the cadence of accountability. It is up to the leader to make sure everyone is focused on improving every two weeks. And only the leader can insist on it. Like a drummer laying down the beat of a song, it is up to the leader to insist that everyone keeps the beat.

The Wheel of Executive Focus

How do you make this happen? How do you lay down the cadence of accountability?

We did it for seven years in a big city, and it worked. Then we repeated the same process for eight years for an entire state government. And it worked there, as well.

It is open source. And this is how.

There are ten business days during a two-week period. With a small, dedicated CitiStat staff of half a dozen people dedicated to preparing and following up on CitiStat meetings, we rotated ten departments through the CitiStat meeting room every two weeks. The CitiStat staff were mostly fresh out of college or new to government—mostly digital natives, eager to learn, and eager to understand, with a couple more senior types. The schedule was set, predictable, and repeatable. Barring holidays or extreme weather events, the meetings always happened.

Every CitiStat meeting is arranged in the same circular fashion. Around one side of the circle are the mayor and his command staff; presenting on the other side of the circle is the department head and their command staff. For one hour, we focus on the operational challenges and opportunities of that single department. When that hour is over, another department comes into the room. And another meeting begins—usually not more than two meetings a morning. And this is how the "wheel of executive focus" rolls through the rotation of two weeks with ten departments having been through the process.

We then repeat the process over the next two weeks, and we never stop.

There is always an agenda, an executive briefing memo, and follow-up memos from each individual meeting, which are written up and distributed to all participants after each meeting. All departments are required to submit their data and any slides they intend to present by noon the day before their meeting.

The process is relentless and demanding, it never stops, and it works.

The goal of this process is not submission—it's empowerment: the building up and nurturing of a collaborative culture—a collaborative culture for solving problems, for making improvements, for making progress, and for lifting up the leaders.

And it requires leadership from the center of the collaborative circle. A leader who says it's okay to ask questions, it's okay to say that something isn't working as well as it could be—a leader who asks for everyone's best thinking.

Policy, Maps, and You

"A policy map is a map where the opportunity to intervene is clear."
—Stephen Goldsmith, former mayor of Indianapolis

People craft policy to effect change. A geographic perspective on policy is always interested in the geographic context of the problem being solved and the solution being implemented through policy. When we make policy maps, we portray a subject in a factual manner to reveal opportunities to intervene and to share any results or outcomes of ongoing interventions.

Mashing up some data onto a map is not a policy map. A dataset of one hundred thousand traffic accidents might be mappable and it might even show which intersections seem to have more accidents than others, but that map is not a policy map. It's a good start, though.

Putting some data onto the map is always a good idea because it starts the process for thinking spatially. The first time that data is shown on the map, questions start to surface about the nature of the data and any patterns that first appear.

Any policy that hopes to address the real issues must be informed by good data and sound reasoning. Most issues that are addressed by policy have a spatial component. Sound policy takes that spatial component into consideration when understanding the problems, reviewing alternatives, implementing new policy, and communicating results.

Bad policy treats symptoms rather than the real problem. Policy maps help to identify the real problem and effect real change.

Putting Together the Puzzle

If you have ever tried to put together a jigsaw puzzle, you know it requires focus and concentration.

You begin by putting all the pieces face up on a table. Then, you start sorting through the pieces in search of patterns of color and contour. If you are working alone, it can take a long time. But assemble a group of people around the same jumble of reality—with a common picture of what it is supposed to look like by the end—and you start making much faster progress. One person focuses on assembling the pieces with straight edges into a frame of the whole picture, another person assembles this color or that color for their own piece of the picture, and so on.

The same is true of solving problems that play out on the map of a city, county, state, or nation. You begin by putting all the pieces of data—your open data—face up on the map for all to see. Then, you pull a group of people around that jumbled reality to identify patterns, to make sense of the various relationships between the pieces.

The GIS map is the table: the common platform for assembling the picture of the latest emerging truth. The data about the problem and its solutions—whether trash, crime, disease, or anything else— are the pieces of the puzzle. Convening the group and focusing its attention on the puzzle is the method of collaborative leadership—the cadence of accountability.

If you keep bringing people together, regularly and consistently, two pictures emerge: the picture of what is; and the picture of what can be—if we choose to do something about it.

Ask More Questions

At one CitiStat meeting early in the development of our own newly established rotation of regular meetings, the Bureau of Solid Waste was at the podium. This critical city service was the primary responsibility of one of the hardest working public servants in Baltimore City government—Joe Kolodziejski, head of the Bureau of Solid Waste.

Top of the agenda for that CitiStat meeting was the problem of unexcused absences. This meant, quite literally, people not showing up for work. Over the years, it had grown into a major problem. And when people don't show up for their shift, others must be paid overtime to complete a collection route or fulfill a given task. It drives costs up, and it also makes us less effective and less efficient.

There were many reasons offered as to why we had such a chronic problem, including high levels of drug addiction and alcoholism among our workforce. Another reason was the gaming of our quick managerial reflex to spend more money to pay overtime rather than fire the chronically absent. Workers get paid time and a half for overtime, so a regular rotation of unexcused absences might just be a way for a crew to boost pay by slowing productivity.

Together, we needed to ask more questions. Not a never-ending range of questions on a host of irrelevant details; not the sort of questions that are designed to make the questioner look smart at a meeting ("Can this scale? Can we take a step back?"). Rather, we needed to ask more questions to get to the root of the problem instead of focusing on the many flowers of its symptoms.

The line of questions and answers on this day unfolded as follows:

Latest premise: "Joe, you told us last week that unexcused absenteeism is one of the biggest operational challenges you face, and you talked about the various reasons why people don't show up for work..."

Question: "When was the last time you wrote someone up (a required precursor to firing) for being repeatedly absent without excuse?"

Answer: "Can't really say, because we don't generally do that."

Question: "Why?"

Answer: "Partly because the union contract has a lot of requirements." (Note: The word "partly" is a clue to ask more questions.)

Next, a short conversation takes place openly around the table because the labor commissioner, the director of human resources, and the city solicitor are all sitting at the table and focused on the same problem. Quickly, we conclude that the union contract requires only some evidence of two or more unexcused absences and one written warning before a dismissal. We agree that we should start tracking the number of times a supervisor writes up an employee for being repeatedly absent without an excuse.

The 5 Whys

Taiichi Ohno, who pioneered the Toyota Production System, is credited with developing the problem-solving technique of the 5 Whys. The core idea is to tie investments directly to the prevention of the most problematic symptoms.

As a problem-solving technique, stopping to ask "Why?" five times gets us to the root cause of the problem and leads to more workable solutions.

Ohno himself gives an example of a machine that stopped working, resulting in an entire production line being shut down for hours.

He first asks, "Why did the machine stop working?" The answer, "Because a fuse blew." But the answer to the first why rarely gets to the deeper understanding of the root cause. So, the next question must be asked, "Why did the fuse blow?" Deeper question, deeper answer. And so the dialogue continues until we finally arrive at the cause of the problem: a five-dollar strainer was left off the machine, allowing metal scraps to get in and wear out the shaft in the pump, which caused the machine to overheat, which in turn caused the fuse to blow, which in turn shut down the machine—and the whole assembly line with it. With the root cause now identified, better preventative maintenance actions can be modified, changes can be made to inspection protocols, and a similar shutdown can be avoided in the future.

The recurring dialogue of the CitiStat and StateStat process gave us the time and focus necessary to ask the 5 Whys.

As you can see, repeating "Why?" five times, like this, can be most enlightening. Answers to the most basic, often overlooked questions can help uncover the root problem and lead to corrective action.

A note is made for the follow-up memo that will go out to everyone after the meeting to memorialize Joe's commitment to track these leading actions of documented evidence and written warning before a disciplinary firing.

That's for the next two weeks. But this is today. And Joe is still at the podium, halfway into our one-hour CitiStat meeting that is focused on the Bureau of Solid Waste. We come back to Joe and the problem at hand.

Question: "Mr. K., you said you don't proceed with firing people who don't show up 'partly because of the union contract.' The union contract doesn't seem to be such a tough requirement, after all, so what was the other reason why we don't fire people for being chronically absent?"

Answer: "Because ultimately, we don't want to lose the permanent PIN number."

Question: "What's a permanent PIN number?"

Answer: "That's essentially the budget authorization that allows you to hire and keep a full-time employee. And if we fire a full-time employee, we can only replace them with a temporary employee."

Question: "Why?"

Answer: "Because that's just the way it is."

Question: "Why is that the way it is?"

Answer: "Because the Finance Department captures all vacant full-time PINs for salary savings, and will only give us a temporary PIN. So, we'd rather have a bad employee in a full-time PIN (with health and retirement benefits) than risk having a worse employee in a temporary PIN."

Okay, now we are getting somewhere. Supervisors don't supervise because they believe their staffing levels will be reduced if they do.

All eyes turn to the finance director—who also has a seat at the CitiStat table—and the next question is asked.

Question: "Do we really discourage managers from holding their employees accountable by taking vacant full-time positions and insisting that they can only be replaced with temps? Or, is this a misunderstanding?"

The finance director believes it is most likely a misunderstanding.

Question: "Are you sure, or is that your best guess right now?"

The finance director pledges to get to the bottom of the question by the next meeting in two weeks—if not before. A note is made by the notetaker, which will also be included in the follow-up memo that goes out to all participants of that day's meeting.

As it turned out, the finance director was able to track down the answer within just a day or two. In fact, on this matter, Joe was right, and the finance director was mostly wrong. At the next meeting, the finance director reports to all in attendance: "This was a rule that was put into effect a few years back—as best as anyone can remember—to close out an operating deficit in the final months of one particular fiscal year. It was not intended to be permanent. We apparently never rescinded the policy."

One mystery solved. One operational barrier to performance dismantled. Joe's supervisors went back to doing an important part of the job of supervising. We tracked the lagging indicators of attendance and performance. We also tracked the leading supervisory actions of writing people up who don't show up for work unexcused. More people started showing up for work, and therefore, more work got done. And it got done with fewer citizen complaints for poor service and fewer dollars wasted on overtime.

In baseball terms, it was a single, not a home run. But with enough singles, you can also put points on the scoreboard.

Building a Collaborative Culture

There are five hallmarks of a collaborative culture—the sort of culture that makes progress possible.

They are the same qualities that good leaders individually demonstrate and practice in their interactions, communications, and relationships with others. This sort of behavior becomes a good contagion. The spirit it creates is like oxygen at the center of the collaborative circle. And it is the practice and demonstration of these qualities that forges the basis of trust, cooperation, and a shared sense of purpose that can ripple through the entire organization over time.

Demonstrating these qualities in the ongoing dialogues around performance measurement and management is key. There is a heightened awareness and attention given to the tone and words of the leader in settings where everyone is watching. Humility is the greatest power, especially in these settings. And the disciplines that follow are all manifestations of the humility of the servant leader, the collaborative leader, and the modern and effective leader.

Inclusion

Good leaders have a default setting of inclusiveness. They bring people into a sense of shared ownership of the challenge and shared ownership of the solutions. They share credit as readily as they share information. They are inclined to set more chairs around the table, not fewer. They foster an atmosphere that is open to new ideas and, most importantly, open to better questions. They invite greater

Working together works. This is one of hundreds of cleanup sites on the citywide volunteer mobilization held every spring and fall. Small things done well make bigger things possible. And the massive citizen mobilizations also allowed us to get ahead of the problem of illegal dumping.

involvement and greater participation of others in figuring out and doing the work at hand.

Acknowledgment

Good leaders acknowledge the effort and hard work of others. They acknowledge the difficulty of another person's job and the sacrifice it requires for the sake of the mission. Good leaders look for opportunities to praise people for a job well done, to thank people for hard work, and to appreciate honesty and candor when the truth is difficult to voice. (My own senior staff took care to highlight for me one or two things worthy of praise in the performance data—especially when they knew I was coming to a CitiStat meeting with an axe to grind. "Pat them on the back for a job well done on hitting the delivery goal on this service, before you hit them for under-delivering on this one that is annoying you." A good CitiStat director will do that.)

Respect

Respect is something deeper than acknowledgement. It has an individual and a collective dimension. There is a reason why societies develop manners over time—things like civility, decency, not speaking over people, and allowing a person to complete their thought before asking the next question.

Communication can be clear without diminishing the mutual respect that is essential for honest discussions and better collaborations. Questions can be pointed without undermining the integrity of working relationships.

Here's an example:

"Bob, you are awful; your team is awful, and you guys are in last place every week."

Versus:

"Bob, I'm sure your team is working hard, but every other team is more productive by any of the primary measures. Is there something the other teams are doing that you guys might try? I know you guys can do better . . ."

It's called respect.

Expect the best, and expect better, and talk to people like they are capable of it.

Recognition

The Irish poet John O'Donohue wrote, "One of the deepest longings of the human soul is to be seen." My father used to tell me that the most beautiful sound to a person's ear is the sound of their own name. This isn't flattery; it is a recognition of human nature and the dignity of the individual person. Know the names of those around the table. Address them by name. Put their names and faces on the slides of charts and numbers that reflect and track the work of their part of the organization.

Congratulate people on the birth of a child, and express proper condolences on the passing of a parent or spouse. Support and visit colleagues when they get sick. This is all basic, decent stuff that, unfortunately, some leaders become convinced they are too busy to do.

Intellectual Curiosity

There is a Japanese management theory of the "5 Whys" (see the sidebar on page 17). It urges those who are in search of the root cause or nub of a problem to patiently ask "why" five times after each successive answer to the prior question. Ask people for their opinion before you give them yours. Ask the people who are doing the job what they believe, think, or feel about an issue or challenge based on their experience. Ask them how they think a process or strategy or tactic could be improved.

Conclusion

Actually, there is no conclusion. Start and don't stop. Lead. Treat people with dignity and respect. Acknowledge good work and lift up the leaders. Foster every day a culture of collaboration. Establish a cadence of accountability with repeatable routines. Bring people together around the latest emerging truth of what is happening and what works.

Progress is a choice.

Begin it now.

Learn & Explore

Data-Driven Government: A New Approach to Governing
Watch a video of a discussion to explore how data-driven decision-making, open data, and performance measurement can impact government policy and effectiveness.

Maryland's GIS Data Catalog
Maryland's GIS Data Catalog is an open data portal where you can search for the latest data from Maryland contributors. The site is part of Maryland's commitment to being open, being transparent, and to provide data that is easy to find, accessible, and usable.

Policy, Maps, and You
View a gallery of maps and apps that are used to describe the current state of things, analyze the opportunities and threats that concern us, and recommend specific actions to achieve shared goals.

Environmental Impact Public Comment App
This app allows the public and other interested parties to directly comment on projects currently being proposed by transportation, environmental, and other agency types.

For links to these and other examples, exercises, and resources, visit SmarterGovernment.com and click chapter 1.

2

When Disaster Strikes

There are really just four questions every leader must be able to answer for their people when disaster strikes: What has happened? What do you know about it? What are you doing about it? And what should my family and I be doing to protect ourselves?

Each is really a question of *where*.

The Baltimore Howard Street Tunnel Fire

I'd spent the day of July 18, 2001, on the Eastern Shore at the annual J. Millard Tawes Crab and Clam Bake—one of the year's major political gatherings. I was still a relatively new mayor, some eighteen months into an intense and totally consuming new job. Every day was full of challenges. Baltimore had more than its fair share of problems to tackle. It was good to have gotten away—to be among the marshy wetlands and rivers of the Eastern Shore, if only for a day.

After spending some time at the hot afternoon festival in Crisfield, it was time to head back to Baltimore. I was just settling in for the long ride home when my phone rang. It was my chief of staff, Michael Enright.

"We've had a bit of a train wreck," he said.

I thought he was speaking figuratively, so I asked how anyone could possibly have screwed up a scheduled press announcement about reduced library hours.

"Not that kind of train wreck—it's an actual train wreck," he said. "Still getting details, but apparently a big freight train derailed and is on fire somewhere in the middle of the Howard Street CSX Tunnel. Turn on your radio."

"Where is the Howard Street CSX Tunnel? And what was the train carrying?"

"It was a freight train, not a passenger train. Not sure yet what kind of freight it was carrying. We are trying to find out."

I asked again, "Can you tell me exactly where it runs and where the tunnel entrances are?"

"One entrance is way up in midtown near the Meyerhoff Symphony Hall, and the other end is right at Camden Yards, where it's currently smoking out the baseball game."

I had never really thought much about the century-old Howard Street Tunnel or where it ran, but my team and I were about to learn more than we ever wanted to know about that mile-and-a-half tunnel. I asked Michael to call me every ten minutes with updates.

As I rode up from the lower Eastern Shore on the far side of the Chesapeake Bay, a black acrid smoke was billowing out of both ends of this tunnel in the center of my city. The southern end was just below Camden Yards, where the Orioles had been playing a doubleheader. The umpires had halted play and fans were nervously making their way through the smoke to their cars. At this point, no one knew exactly what kind of chemicals were on that long train or which ones were in danger of catching fire next.

Minutes after I hung up with Michael, the police commissioner called to talk with me about the pros, cons, and parameters of a possible evacuation. Then my press secretary called, because he was getting calls from the press about whether there were evacuation instructions.

As I talked by phone and listened to a scratchy signal from WBAL radio, I could hear the tension in the reporter's voice. The smoke was thickening and so was the mystery. City air raid sirens began to wail in the background.

"What does that mean?" the reporter wondered aloud. And so did I.

"Who the hell ordered the air raid signal without telling people why?" I grumbled in the car, wondering whether a more serious public health emergency than fire might be coming to a boil as we urgently sought answers.

After twenty anxious minutes, the radio newscaster was able to pass on official instructions from the fire department. People attending the Orioles game at Camden Yards needed to leave. People living in the center of town, at the northern end of the tunnel, needed to "shelter in place." Apparently, the fire department had beaten me not only to the air raid siren but to the evacuation decision, as well.

That was better, I supposed, than no decision; better than no instructions at all.

I finally made it into town at 6 p.m. and dashed to the police command center where I had asked everyone to huddle. In hindsight, I should have asked to meet my senior staff at fire headquarters—but I was relatively new to the job of mayor, and not yet very adept at asking the question "where?" There in the CompStat Room of police headquarters, police and fire officials were plotting road closures on paper maps spread across desks and sometimes projected onto large screens.

They were also communicating with first responders on the scene as the fire grew to five alarms. Because I had called this command huddle in police headquarters, there was no map we could see that told us where fire apparatus were being deployed. By early evening, the police commissioner advised that we should close the major interstate into the city. And this city decision now required some additional coordination with State Highways—who were presumably operating out of their own space somewhere looking at yet another paper map.

As I paced the police command center, word came that the intersection of Howard and Lombard Streets, above the tunnel, had suddenly become a raging white-water rapid. A giant water main had become so super-heated by the tunnel inferno below that it ruptured and cracked—splitting the asphalt above and sending a flood of water down the streets. This was a major intersection of our downtown for car traffic and for mass transit. The violent upheaval of the street cut the north-south light rail line in half. Cars were coursing like rubber rafts, businesses were flooded, and pedestrians were running for high ground. And to make matters worse, there was no water pressure now in the heart of downtown— where the fire department needed it to put out the fire.

This was fast becoming a bigger and longer-lasting emergency. And so much was still unknown.

Black smoke billows from the Howard Street Tunnel outside Oriole Park, Camden Yards, Baltimore, on July 18, 2001, after a freight train carrying hazardous materials derailed. Clearly, there was a fire somewhere inside the 1.8 mile long tunnel. The question was, where?

Baltimore City firefighters extinguish a still-burning rail car that was pulled from the inferno of the Howard Street Tunnel days after the emergency first began.

Where exactly was the train? Was it still connected or was it in pieces? Where were the pieces? Where was the actual fire burning and how close was it to people? Where was it in relation to other power and water infrastructure? What chemicals were on the train? Where were those chemical cars in relation to the cars that were on fire? Could there be an explosion? How close could our firefighters and police safely get to the train? How close should they get?

Firefighters tried repeatedly to enter the tunnel from either end but were pushed back by that smoke and heat that reached 1,800 degrees Fahrenheit.

Some initial air testing had been done. So far the smoke—other than being smoke—was not a more toxic or lethal gas mixture. Beyond that, there was not a lot to say except to advise citizens to shelter in place—stay indoors. But, I knew people needed to see and hear their mayor saying it.

We decided to address a gaggle of reporters and cameras outside Meyerhoff Symphony Hall, where we could point to the air monitoring now being done at the northern entrance of the tunnel where smoke still billowed. This was the end of the tunnel where the greatest numbers of people lived, as well—mostly in high-rise apartment buildings.

As a hazy dusk settled on the city, I went on the air to tell people what happened, what we knew, what we were doing about it, and what they should do to protect themselves and stay out of the way of the first responders.

It was not a very satisfying or reassuring experience for the press, the public, or for me. We decided—especially given the pace of cascading events and lack of situational information—that we should brief the media every hour after that.

An hour later, after one more unsatisfying and confusing late-night press conference, we made another important decision: the next press briefing, and every press briefing thereafter, would have a map of the incident front and center. By dawn's early light, thanks to our twenty-something graphics guy, Frank Perrelli, we had our map. Rectangles cut out of Velcro or felt to represent box cars and chemical cars. Colored markers. Refrigerator magnets glued to cardboard firefighter symbols. It wasn't pretty, but it did the trick. And it was the best Frank could do after breaking into offices in City Hall at 3 a.m.

Unified Command and Situational Awareness

During the Howard Street Tunnel fire, I was getting a crash course (or, more accurately, a train crash course) on some key concepts in emergency management and leadership—concepts I really wished someone had taught me before then.

In the initial fog of the unfolding emergency, I listened as some junior fire officers complained and grumbled about the lack of a "unified command." The third time I heard this term used was when a fire captain—in my presence—told the fire chief, "With all due respect, sir, we need to quickly establish a

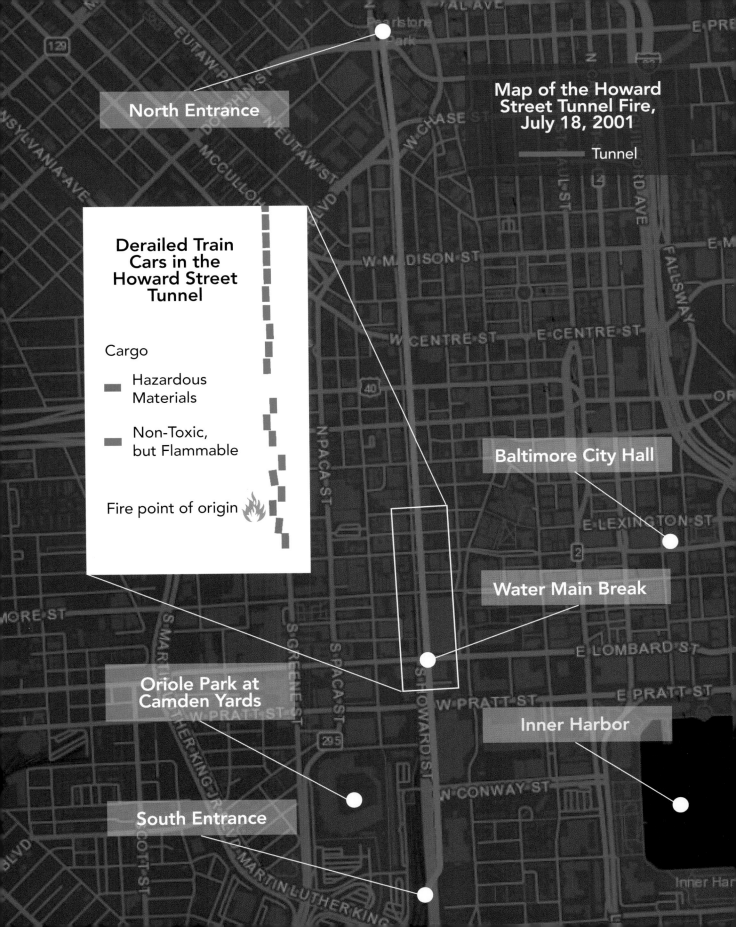

Map of the Howard Street Tunnel Fire, July 18, 2001

—— Tunnel

North Entrance

Derailed Train Cars in the Howard Street Tunnel

Cargo

—— Hazardous Materials

—— Non-Toxic, but Flammable

Fire point of origin 🔥

Baltimore City Hall

Water Main Break

Oriole Park at Camden Yards

South Entrance

Inner Harbor

Data Drills

Preparing for Disaster with "Data at the Speed of Thought"

In a city, an analytics project usually begins with a question. But when we respond to an emergency, we need to be prepared well before the question arises.

In New York City, we created a concept called "data drills." A data drill is a multi-organization collaboration exercise that is used to gain greater insight into how a city collectively thinks about, manages, shares, and uses data. During a drill, we discover data and build relationships, so people who are responding to an emergency have access to "data at the speed of thought."

Generally, data drills are developed and conducted based on some operational challenge that involves data and which will require multi-organizational cooperation to achieve a desired result. Drills can be designed for (but are not limited to):

• Specific scenarios (hurricane flood zones, homeless counts, data center disruption)
• Capacity building (collecting data, learning how to operationalize a specific dataset)
• Operations development (two agencies cleaning up downed trees in a joint operation)
• Testing software (testing new features in a data-sharing platform)

Data drills are meant to help us take on that challenge by having organizations across the city surfacing, sharing, and integrating data. A drill takes place over a designated period, with a specified start time and end time. It forces all the participants to work within similar time constraints that we tend to see in real life. Every data drill results in overall city-wide data IQ growing ever so slightly.

Data drill deliverables should be defined early in the planning phase. They might include (but are not limited to):

• Identification of datasets, with metadata and data dictionaries
• An organization-specific operational workflow relevant to data and use-cases
• Interagency workflow for operations, analysis, and network infrastructure
• List of organization contacts, roles, and responsibilities
• Documentation of activities and observations
• Report, with recommendations

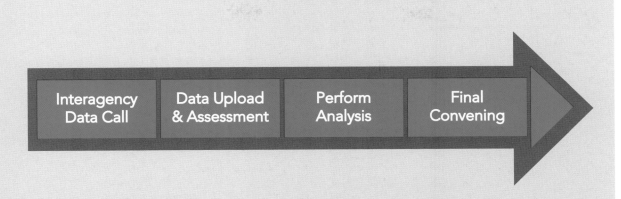

Why did we decide to do data drills in New York City? We realized that even when we weren't thinking about using analytics to solve a problem, we needed to be thinking about data all the time.

Former Secretary of Defense Donald Rumsfeld famously said that there are "known knowns," "known unknowns," and "unknown unknowns." We understood that when it comes to data, this quote is extremely relevant. There is data that we know we have (known knowns), there is data that we know we don't have (known unknowns), and there is data that we know absolutely nothing about, including even the fact that it exists (unknown unknowns). After 9/11, Hurricane Sandy, and the Legionnaires' disease outbreak in New York City during the summer of 2015, we knew very well that it's the unknown unknowns that hurt you the most.

Data drills are a mechanism for helping a city to baseline where they are with citywide data practices. They're also a mechanism for helping a city improve its ability to identify, understand, and use data to solve a city challenge when requested.

Data drills make the city smarter about its data, and that is key to being able to use data and analytics to make a city safer, smarter, healthier, more efficient, resilient, sustainable, and equitable.

Regardless of whether urban analytics are immediately necessary to remediate a situation, for any city, data drills should be considered phase zero—constantly running in the background at a cadence that keeps the city's data ready to be put into action.

—Dr. Amen Ra Mashariki, former chief analytics officer for the City of New York

Baltimore City firefighters at the north end of the Howard Street Tunnel in the first hours of an emergency that would shut down our entire Central Business District for days.

unified command." And with that, he glared directly at me.

Light bulb—I needed to figure out what is meant by a "unified command." And I needed to figure it out quickly. This is what I learned on the job that day.

There is a concept of operations in emergency management called the incident command system (ICS). It is a standardized approach to the command, control, and coordination of emergency response.

Here's a simple example of how this approach works: In a smaller incident, like a house fire, the fire truck shows up. The officer in charge of putting out the fire commands the personnel on the fire truck. Lives are evacuated or rescued, working hoses are attached to working fire hydrants, and the fire is extinguished. If assistance from other departments is needed to close a street or find temporary housing for a displaced family, the incident commander directly calls up the help from his peers in other departments in a routine way. There is no need to go to the head of transportation or housing or to wake up the mayor in the middle of the night. It is routine. It is a practice that has been operationalized in daily events over the course of years and decades.

Such routine fires are handled in a classic triangle, if you will, of command and control. The operational incident commander is at the top of the top of the triangle. Orders flow down through a command structure. Things get done on the basis of the commander's orders.

But when an emergency like the Baltimore Howard Street Tunnel Fire grows in scale, complexity, and the time and resources necessary to address the emergency, a different type of command organization becomes necessary. This is what the Federal Emergency Management Agency (FEMA) training manuals call a "unified command."

The triangle of incident command and control grows in size. And it is augmented at the top of the triangle by a collaborative circle—a circle that integrates the otherwise independent and separate departments of government into a focused supporting role to the leadership of the primary incident commander. The unified command thereby becomes a single, coordinated circle of command support. Not a tactical or operational decision-making body, but a coordinated command in support of the operational decisions being made by the lead department—in this case, the Fire Department—whose primary responsibility is to first extinguish the fire.

A unified command features a single, integrated incident organization; one shared understanding—around a single map—of incident objectives and the sequence of actions necessary to achieve them; one operations section, with one chief of operations being supported by incident chiefs of other departments, and a coordinated process for ordering and deploying resources to bring the emergency under control. At the center is the map, a common frame of reference helps all emergency management and public safety personnel maintain situational awareness during an incident.

As a newly elected mayor in pre-9/11 America, I had not been trained or even been given a primer to understand any of this. And regrettably, some of the early decisions I made—like asking the chief of the Fire Department to meet me at police headquarters—certainly proved my civilian ignorance to our first responders.

But I would quickly learn that my role for the next few days was to become the best intragovernmental deputy that any fire chief ever had. My role was to enforce the co-location, regular communication, and convening of the highest levels of departmental leadership to focus on the latest emerging truth. My job was to keep that collaborative circle focused on supporting the efforts of the fire chief and his operational incident commander.

It was also my job to make sure we were telling our citizens—on a regular, recurring basis—what happened, what we knew, what we were doing about it, and what they should be doing to protect themselves and their families.

Cascading effects: An official from the city Bureau of Water and Wastewater surveys the break in a 40-inch water main caused by a fire in the Howard Street Tunnel directly below it. The break turned a major intersection into whitewater rapids, and it shut down water services to every building and business in the central business district for days.

Putting Out the Fire and Getting Back to Normal

By nightfall on the day the Howard Street Tunnel Fire started, we had finally established a unified command. And until this fire was extinguished, everyone in our administration understood that this was a fire incident; therefore, the fire chief was in charge. Police, Water and Wastewater, Public Works, Traffic—they all now supported and, as a practical matter, reported to the fire chief. I stayed a step back from the chief on the ground. My chief of staff, Michael Enright, and I took alternate twelve-hour shifts, with mine timed around early morning and end-of-the-day press briefings.

As the operation progressed, a clearer picture came into view for all of us. As our firefighters made progress, we were able to convey that progress to the public. CSX finally provided us with the manifest. Maddeningly, the manifest told us that there were some potentially lethal chemical cars, but CSX could not guarantee that the cars were in the order listed. We were going to have to figure out where the most dangerous and toxic cars were in the jumbled, smoldering, underground inferno.

The map Frank Perrelli had created became the central organizing tool for integrating the work of many departments in support of the fire chief. The map also allowed us to tell the story to the public—with better information at each successive press briefing. We answered the questions of where things were happening and what we knew about them and explained what actions we were taking where. We steadily filled in the missing information on the map as we learned more and brought the emergency under control.

OSPREY

A Common Platform for Emergency Management and Preparedness

Craig Fugate, former administrator of FEMA, once quipped to me, "The problem with common operating platforms is that everybody wants to keep their own..." Usually turf objections (masked as interoperability issues) are enough to keep most governments from actually creating a common operating platform—even for emergency management.

The State of Maryland, with Esri's help and with some great work by the Center for GIS at Towson University in Maryland, created its own common operating platform for emergency management. And because every such platform gains credibility by having an indigenous animal-spirit acronym, we called ours OSPREY, after the high-flying, far-seeing, fish-eating hawk that inhabits the Chesapeake Bay during warmer months of the year. OSPREY stands for Operational Situational Picture for Response to an Emergency.

OSPREY helps us better manage large-scale emergencies, and the recovery operations that follow. And it gives citizens a view into the unfolding nature of the emergency and the recovery, so moms and dads know what they should be doing to keep their own families and businesses safe.

Common operating platforms like OSPREY not only help us see and model dynamic events, they also increase our ability to deploy resources and take lifesaving actions more quickly. These new capacities were enabled for the first time in human history by the new technologies of GIS and the internet.

OSPREY's map-based common operating picture enables first responders to seamlessly share critical static and dynamic emergency management information in real time. OSPREY compiles data from many sources to give a composite picture of Maryland-specific data and hazard-related information and to answer the questions "What's happening around me?" And "What can I do about it?"

At a glance, the OSPREY dashboard provides a snapshot of what's going on in Maryland by region. The status of power outages, weather, traffic, hospitals, and shelters are color-coded in an easy-to-digest table. Automated alerts include power, weather, traffic, hospitals, and shelters. A mouse click or tap on an alert box provides the public with additional details about alerts and relevant links.

OSPREY was developed by the Center for GIS at Towson University using the ArcGIS® platform.

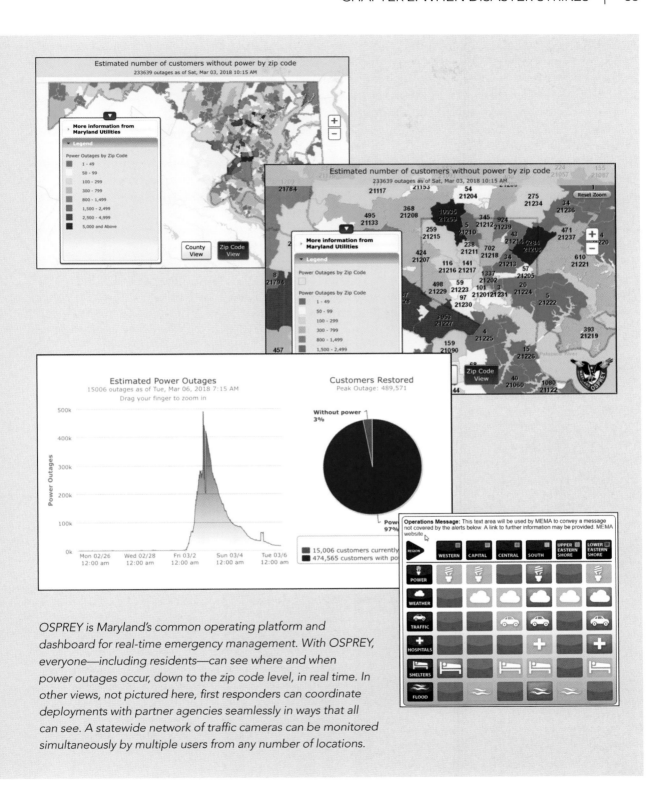

OSPREY is Maryland's common operating platform and dashboard for real-time emergency management. With OSPREY, everyone—including residents—can see where and when power outages occur, down to the zip code level, in real time. In other views, not pictured here, first responders can coordinate deployments with partner agencies seamlessly in ways that all can see. A statewide network of traffic cameras can be monitored simultaneously by multiple users from any number of locations.

Our downtown would be closed for nearly a week. More than 1,200 buildings lost power. Local bus and commuter train service was suspended. Internet and cargo train service was disrupted all up and down the East Coast. We had to close the Inner Harbor to boat traffic and inform an unhappy Orioles management that baseball games—ultimately, three games—must be canceled until further notice.

Five days after it started, the fire was finally extinguished. It turns out, the giant, ruptured water main burst right above the most intense part of the fire. In hindsight, it had served as a giant sprinkler system—albeit, a costly one.

As the charred and most toxic chemical car creaked through Mount Royal Station around midnight on day five of the emergency, a crowd of exhausted firefighters said many a silent prayer of thanks. The tunnel was a mess, and the streets above were badly damaged. Their repair, and the repair of the water main and the light rail line, would take many weeks. But our firefighters had managed to contain the complex fire. The most toxic chemicals remained confined in their soot-blackened containers. And not a single life had been lost.

I learned some important lessons in emergency management over the course of that week, among them: The importance of establishing a unified command. The vital necessity of communicating regularly to the press and to the public. I also learned the value of something we should have had before the emergency struck: a common operating platform—a shared map that would allow us to understand what was happening where.

In that July of 2001, our country considered the Baltimore Howard Street Tunnel Fire a big, unprecedented city emergency. But that was two months before the attacks of September 11. And after that fateful day, most other emergencies now seem small by comparison.

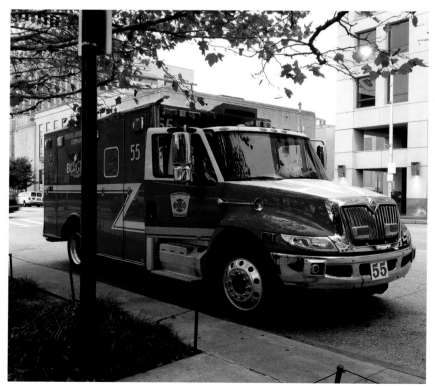

A commemorative manhole cover (above) designed by ace city graphic artist and webmaster, Frank Perrelli. When recovery and reconstruction were finished, we placed it in honor of our firefighters in the street above the tunnel—over the location where they brought flaming cars under control. A Baltimore City Fire Department truck (left).

The Day the World Changed

One bright fall morning in September, my mother, my brother Peter, and I were traveling to New York City—the borough of Queens, more exactly—where my younger brother Patrick was running for city council. My father and other family members had been campaigning there all weekend. When an O'Malley is on the ballot, the tribe rallies.

Peter was driving. I had told my police executive protection detail to stay behind, because I'd be working a polling place out of state all day. As we reached the New Jersey Turnpike, one of the detectives from the executive protection detail called. "Sir, you might want to turn the radio on. They think a plane crashed into a big building in Manhattan. There will be a lot of traffic today in Manhattan, so you might want to go around it."

I flashed back to the day a small plane had landed on the upper deck of Baltimore's Memorial Stadium after a Colts' game in 1976.

"Was it a small plane, like a Cessna?" I asked.

"They don't know," he responded. "But on TV, there's a big hole and a lot of smoke. Lord knows how they are going to get the fire out that high up."

We pulled into a rest stop and grabbed breakfast to go. As we paid the cashier, we watched the image of that big smoking hole on the restaurant's television. We got back in the car, still not realizing that the world had changed. Then, merging back into highway traffic, we turned on the radio: A second plane, a large passenger airliner, had hit the second tower of the World Trade Center.

I was riding up front and turned to Peter. I didn't have to say, "Turn around and get me back to Baltimore." He knew instinctively to make a quick U-turn across the emergency vehicle break in the median.

My mother wanted to check on our dad and sister in New York. But overwhelmed by callers with the same urge to connect with loved ones, cell phone service began to fail. Soon, I couldn't get through to anyone in my own city hall.

My old numeric pager started going off, first with the police commissioner's number, and then with the deputy mayor's. Then, the radio announcer reported that a third plane had crashed into the Pentagon. The entire continental United States had been declared a no-fly zone.

I prayed to God to protect my city until I could get back home.

My detail, along with a motorcycle escort, met me on the side of the highway at the Baltimore County border. When I got to the city, I went directly to the police commissioner's office. Tactical officers flanked the entrance of police headquarters in full SWAT mode and armed with submachine guns. It was the middle of a weekday afternoon, but the city seemed eerily deserted. Everyone who could had gone home to be with their spouses and kids.

Police Commissioner Ed Norris, who had come to Baltimore after a full career in the New York Police Department, sat at his desk, staring at the live television reports from his beloved New York. Over and over again, he kept saying, "I can't believe they brought down the towers."

After a quick operational rundown, we held a press briefing to let local media know what actions we were taking in our own city.

"Were the SWAT teams and submachine guns at the entrances to city buildings and Baltimore's World Trade Center a bit of overreaction?" asked a skeptical reporter. Numb but not dumb, Commissioner Norris shot back, "I dunno. What do you think?"

There was no playbook for this new reality.

My wife Katie called to tell me I needed to come home so William, our three-year-old, could see that I

An "All Crimes, All Hazards" Approach

After the attacks of 9/11, my administrations—in both the City of Baltimore and the State of Maryland—strived to make us the national leaders among cities and states in homeland security and preparedness.

We intentionally adopted an "all crimes, all hazards" approach, which means we developed core capacities that are of use and application every day in fighting crime and responding to emergencies. In other words, if we could track down the license plate of an elderly citizen with Alzheimer's who has gone driving off, lost and missing, we would be better able to track down the license plate of a terror suspect before or after an attack. If we could maintain interoperable radio communications among all first responders statewide during a hurricane, we could also maintain interoperable statewide communications in the aftermath of a terror attack. And so on. The Maryland Coordination and Analysis Center (MCAC), our fusion center for intelligence gathering, analysis, and dissemination, likewise was greatly improved and became a national leader.

Under our administration, the foundation of Maryland's homeland security program was twelve core goals that represented fundamental core capacities critical for effective response to large-scale natural or man-made emergencies. Progress toward the twelve core goals was measured individually through performance metrics developed by the Governor's Office of Homeland Security for monthly homeland security-stat meetings. As time progressed, we added a thirteenth core capacity—cybersecurity.

Goal 1: Interoperable Communications

We initiated construction of Maryland's first statewide interoperable radio system, providing coverage to more than 55 percent of the population. We also built the state's first interoperable computer-aided dispatch (CAD) and automated records management systems (RMS) network and deployed OSPREY, a GIS-based common operating picture for homeland security and emergency management.

Goal 2: Intelligence/Information Sharing

Building on years of reform, including transitioning the MCAC from terrorism-only to all crime; doubling Maryland State Police personnel; physical rebuilding of the space; and creating new daily gun, homicide, and intelligence reports, we received back-to-back perfect scores of 100 on the Department of Homeland Security's national fusion center assessment in 2013 and 2014. We also provided the resources to network almost 80 percent of license plate readers across the state into a central database.

Goal 3: Hazmat/Explosive Device Response
We expanded Type 1 hazmat and bomb squads to every region in the state, improved response time to one hour (four times better than the national minimum standard), and provided millions of dollars in new equipment for defending against explosive improvised devices (IEDs), including an entirely new maritime team. We also created a new nationally recognized comprehensive statewide Preventive Radiological and Nuclear Detection (PRND) program, with participants from more than eighty federal, state, and local first responder agencies. We also became the first state in the nation to sign a memorandum of understanding with the National Nuclear Security Administration to improve detection, prevention, and response.

Goal 4: Personal Protective Equipment for First Responders
We provided more than two thousand full sets of interoperable personal protective equipment (chemical suit, air mask, and other equipment), so every Maryland tactical and patrol officer in the state's five largest police agencies can protect themselves and assist the public during a weapons of mass destruction (WMD) attack or pandemic.

Goal 5: Biosurveillance
We expanded the state's biosurveillance system from only seventeen hospitals to all forty-six acute care hospitals, absentee data from all twenty-four Maryland public school systems, and real-time, over-the-counter drug sales data from more than three hundred pharmacies, covering all twenty-four counties in the state.

Goal 6: Vulnerability Assessment
We set the goal that every region in the state should have a comprehensive all-hazards threat and vulnerability assessment in place, including an assessment and inventory of critical infrastructure in the region. They should be fully updated every three years. We created a complete inventory of critical infrastructure, including assets controlled by the private sector, and other potential targets, such as communities and populations of interest. This inventory included a regularly updated assessment of specific vulnerabilities that identified major gaps where funds should be invested to harden the most vulnerable and at-risk targets.

Goal 7: Training and Exercises
We established quarterly governor/cabinet-level program exercises focused on hazards, from hurricanes to cyberattacks. We rebuilt the after-action program of the Maryland Emergency Management Agency (MEMA), identifying and following through on almost 150 specific corrective-action items for large events such as the 2012 North American derecho and Hurricane Sandy.

Goal 8: Closed-Circuit Television (CCTV)

We created MVIEW, a common operating platform for CCTV that is accessible at command centers and in the field on tablets and smartphones. We increased adoption from fewer than two hundred traffic cameras in 2007 to more than two thousand traffic, security, and vehicle-based cameras and 1,676 users in 2014, networking previously incompatible systems.

Goal 9: Mass Casualty/Hospital Surge

We received a perfect score of 100 in public health preparedness from the Centers for Disease Control and Prevention's annual assessment for three years in a row, increasing from a score of only sixty-seven in 2008.

Goal 10: Planning

We completely reformed statewide emergency planning by creating the Maryland Emergency Preparedness Program (MEPP)—a framework for coordinating and organizing state emergency response plans and creating a continuous revision cycle that aligns risk assessments, federal grant funding, and the acquisition of resources. We also invested more than $2 million to create a state-of-the-art State Emergency Operations Center (SEOC) with refreshed technology. We provided better situational awareness with an innovative video wall that displays MVIEW, Maryland's CCTV statewide network.

Goal 11: Backup Power and Communications

We created a comprehensive inventory of state and local backup power generators for emergency managers, built a real-time, publicly available power outage map, and passed rules to require new school construction to include backup power to increase the number of local shelters with backup power. We also installed 168 Digital Emergency Medical Services Telephones (DEMSTEL), or Voice over Internet Protocol (VoIP) communications, to provide backup communications between critical facilities such as hospitals, health departments, police barracks, and 911 call centers.

Goal 12: Transportation Security

We implemented nation-leading port security and airport security, created the Maritime Law Enforcement Information Network (MLEIN) radar-based surveillance system of the Chesapeake Bay to manage security and natural resource conservation, created common platforms for public safety (including MD First, CAD/RMS, OSPREY, MVIEW, ESSENCE, license plate readers, and MLEIN), and implemented MEPP. MEMA completely reformed not only the physical space of the SEOC, but the way it conducted itself and the way planning and coordination was done in Maryland.

Goal 13: Cybersecurity

Tapping into the expertise of our own National Guard in cybersecurity, we began to test, game, and improve upon the cyber defenses and response capacities of the various departments of our state government. We launched a statewide cybersecurity training program that targeted more than 43,000 state employees and achieved 90 percent completion of monthly training modules. In addition, we created the position of cyber security director and initiated cyber-vulnerability assessments of all major state agencies to help protect citizens' personally identifiable information.

By the time our service in the governor's office was over in 2015, Maryland was at the forefront of a new preparedness curve, and Maryland's leadership was widely recognized by our peers. There was never a day when we felt we had done all we could. And every day, we worked to continually improve these capacities.

Population Covered by Statewide Interoperable Radio	Hospital Emergency Department Visits Covered by ESSENCE	Traffic Management Video Feeds Available to First Responders
		2014 1,657 Video Feeds
2014 45%	2014 100%	2009 266 Video Feeds
2006 0%	2006 40%	
45.0% INCREASE	60.0% INCREASE	522.9% INCREASE

Strategic goals and dashboards: **The more important the goal, the more important it is to measure progress.** *This public-facing dashboard shows progress (as of July 2, 2014) toward the goal of delivering and maintaining Maryland's initial twelve core capacities for homeland security preparedness by 2016.*

was okay. He feared the falling buildings on television were in *his* downtown, where his father worked. He was crying. She put him on the phone and I did my best to not sound worried.

I vowed that day that we were going to become the best prepared city in America against this new threat. And I had never been more serious about anything in my life.

Given the triangulation of the attacks north, south, and west of Baltimore, it felt very much in the weeks after 9/11 that a second attack was imminent. It was not an unhealthy mindset. I had a sobering conversation with the city council president and council vice president about civil succession.

On September 12, I could not get anyone in official Washington to return my calls to tell me what I should be doing to protect my city. So, I called former senator Gary Hart, upon whose presidential campaigns I had worked in my even younger days. I knew Hart had co-chaired the important but little-noticed bipartisan US Commission on National Security/21st Century, chartered in 1998. In the commission report, released in January 2001, Hart had predicted the likelihood of terror attacks on American soil with great losses of American lives. The commission called for broad changes in security policy and actions. Those warnings and their recommendations had tragically gone unheeded.

"Why are you calling Washington?" Hart asked. "They will be forty years catching up with this new reality. You cannot wait for the federal government, Mr. Mayor. Your people need you to act now."

He continued, "Surround yourself with the smartest people you have—from Johns Hopkins University to Aberdeen Proving Ground to the National Institutes of Health—and put together your own Baltimore security council, your own plan, and do it now. When these guys come back for the next attacks—and they will be back—you want your city numbered among the best prepared. Harden targets, up your game plans, and work toward resilience. But don't wait on Washington."

It was good advice. And we took it.

When this new Baltimore security council was quickly called together, we started asking questions. What were our greatest vulnerabilities? Where were they? Who was responsible for securing them? Did we need to guard our reservoirs? Did we have supplies and plans in place to inoculate our population

The Tribute in Light—Two vertical columns of light in place of the Twin Towers, in remembrance of the victims of the September 11, 2001, attacks.

during a biological attack? Did our private-sector partners—who maintain our energy grid, the rail system, and the port—know their vulnerabilities? Were they acting to protect them? Did our first responders have the protective gear they would need to safely perform their duties during a biological or chemical attack? Could they communicate with one another during a large-scale emergency—or would they suffer the fate of the firefighters who climbed to their deaths in the World Trade Center, even as police were being ordered on their radios to evacuate?

I checked in regularly with Tom Cochran, executive director of the US Conference of Mayors, to learn what other mayors were doing. Given our proximity to Washington, DC, Tom asked me to host some video conference calls to share information and strategies as mayors across the country phoned in to figure out what to do in the face of the new reality.

The country was grieving, but united. And in cities across America, people were looking to their mayors for leadership.

Preparedness in the face of twenty-first-century threats remains a work very much in progress today. (See the sidebar "An 'All Crimes, All Hazards' Approach.")

And that work also calls for smarter government.

Learn & Explore

OSPREY Dashboard
Maryland Emergency Management Agency's OSPREY system brings together data from multiple automated sources in near real time.

Virtual Hurricane Resource Center
The City of Chicago's Virtual Hurricane Resource Center is your link to community and government-based services and resources available from the city, the State of Illinois, and federal agencies.

Are You Ready, Tampa Bay?
This story map can be used by citizens who want to know more about what vulnerabilities the community faces and how the city plans on mitigating risks.

For links to these and other examples, exercises, and resources, visit SmarterGovernment.com and click chapter 2.

To better deploy recovery efforts, this hurricane damage viewer app by Esri lets you compare before (left) and after (right) images of areas in Florida impacted by Hurricane Michael by swiping the center bar across the image on-screen.

Collaborative Leadership

In all human endeavors, leadership is the great variable. But the elements of effective leadership are no mystery. It's a practice—a practice of repeatable routines. We have never had better technologies for measuring and mapping performance. We have never had better tools for modeling our built environments and our natural environments.

What we lack isn't information or technology; it's leadership.

Bringing People Together to Get Things Done

I used to think effective self-governance was 10 percent policy and 90 percent follow-through. I was way off. Through teaching and reflection, I realize now for the first time—or perhaps remember for the first time—that today governing is 10 percent policy, 40 percent follow-through, and 50 percent leadership.

We live firmly in the Information Age. So much information, so much opportunity for progress, and such a crucial need for collaborative leaders—leaders with the ability to *bring people together to get things done.*

This book presents a lot of great examples to assist and inspire you on your own journey, but progress is not a simple paint-by-numbers exercise. Progress is a collaborative enterprise. There is an alchemy required—especially in the context of the Information Age—for maintaining the precious consensus necessary to get difficult things done. It requires a human element more powerful than technology or management prowess. That element is called leadership.

As one of my own college students observed, "You can have all the great technology you want, but without leadership, it doesn't amount to a hill of beans."

So very true—A+.

Two Essential Questions

When I was elected mayor in 1999, my city, Baltimore, had become the most violent, addicted, and abandoned city in America. At one of our first community meetings in a hard-hit neighborhood of East Baltimore, citizens assembled with me, their new mayor, to talk about crime, public safety, justice, and racial injustice—no easy topics.

There was tension in the air. A fear about what might be.

A little girl came up to the microphone.

"Mr. Mayor," she said, "my name is Amber, and I am twelve years old. And because of all the addicted people and drug dealers in my neighborhood, there are people in the newspaper who call my neighborhood 'Zombie Land.' And I want to know: Do you know that they call my neighborhood Zombie Land? And are you doing anything about it?"

Her questions are the two essential questions for all leaders today:

Do you know? Are you doing something about it?

The questions she asked were really questions she's asking of all of us. Behind all our data, there are people—living their lives, shouldering their struggles—who deserve a government that works.

The great promise of effective governance in the Information Age is not so much that the data allows us to manage the masses, but rather, it allows us—if we care—to see the needs and dignity of every individual person.

It's not about the technology.

It's about our relationships.

It's about knowing, and about caring.

It's about seeing and understanding the connections between people, places, and things.

And it's about leadership.

The New Rules of Leadership

Western democracies have some catching up to do with consumer expectations. According to a recent study completed by the Pew Research Center, 65 percent of Americans go online to find information they need about their government, but only 10 percent report finding the information they need.

If Amazon, Uber, and a host of other companies can provide better service thanks to the new technologies of the Information Age, why can't our governments? If the GPS system in my car can navigate me to the quickest route through traffic congestion and fender benders, why can't my government use these same technologies to better anticipate these routine accidents?

Technology isn't the problem. The technology is proven. Nor is cost a barrier; the availability of these new technologies is widespread and relatively inexpensive. The problem is the great human variable of leadership.

Old habits die hard. But a new way of leading and governing is emerging.

The streets of Baltimore. Established in 1763, Fell's Point is one of the oldest neighborhoods in Baltimore. It features cobblestone streets, a vibrant public market, waterfront restaurants, crusty old bars with nautical names, and a Revolutionary history that is second to none.

In Silicon Valley, people who keep trying new things—even though they sometimes fail—are called innovators and entrepreneurs. In government, people who try new things and fail are fired or voted out of office. Therefore, public administration has developed a very slow, cautious, and risk-averse approach to embracing new technologies. Over the course of time, this has created a tyranny of "the way we have always done it" in public service.

But in most big cities across America today, call centers and customer service guarantees have become the new normal. In some places, historic data is being combined with predictive analytics to pre-deploy tow trucks to the places where minor accidents most frequently happen at rush-hour, or to deploy police patrol cars to the tiny squares on the map where crime most frequently happened during the same eight-hour shift over the prior ten years.

A new generation of leadership is changing the old mindset. And a sharp contrast is emerging between old and new.

The old way of leadership was characterized by closed structures—hierarchy, bureaucracy, command and control, and information tightly controlled at the top. The new way of leadership is characterized by open structures—common platforms for collaboration, open data, and timely, accurate information shared by all.

This new way of leadership is entrepreneurial, performance-measured, and interactive. Authority is increasingly based not on the old law of "because I told you to do it," but on the new law of "because I can show you it works."

This new way of leadership is not effortless. It requires work. It requires a different kind of discipline. Most of all, it requires a relentless commitment to bringing people together in short regular meetings focused on the latest emerging truth of what is—truth about conditions on the ground and the actions being taken to change those conditions. And the purpose of laying down a repeatable pattern of recurring meetings is not simply to have meetings, or to "ooh" and "ah" at pretty maps, but to figure out better ways to coordinate, communicate, and cooperate to produce better results.

This new way of leadership also requires will at the center of the collaborative endeavor. The will—and the courage—to follow the evidence wherever it might lead. The will to try new things to see if they work to deliver better results. But first, it requires the will to begin, and the will to persist.

Therefore, the first rule of this new way of leadership is simple: **Start and don't stop**.

Rule #1: Start and Don't Stop

Most people fail at standing up a performance management system by simply deciding not to start. Good intentions get overwhelmed by the crisis of the day. Plans to implement a performance management regimen get pushed to the back burner by the pressing problems of the moment—the inauguration, the transition, the midterm elections, re-election.

Somehow, the need to stand up new systems never seems as urgent as the need to get through the week or the day.

I once met with the newly elected mayor of a city who had inherited some great tools from his predecessor for measuring performance and delivering better service. But the newly elected mayor had convinced himself that a big fiscal fix needed to be accomplished before he could start any regimen of regular performance management meetings. As every super-busy day passed, so too did the golden time of transition—that short period of "new beginning" that every newly elected administration must institutionalize new systems and new practices.

In some places that golden time can last a whole year; in other places, just a few short months. The average is about six months. As cleared fields naturally revert to forest, big bureaucracies naturally revert to "the way we've always done it." The field of dreams can quickly become overgrown by a culture of excuses.

In one of my favorite quotes about the spark of human creativity, William Hutchison Murray writes:
"Until one is committed, there is hesitancy, the chance to draw back, always ineffectiveness. Concerning all acts of initiative (and creation), there is one elementary truth, the ignorance of which kills countless ideas and splendid plans: that the moment one definitely commits oneself, then Providence moves too. All sorts of things occur to help one that would never otherwise have occurred. A whole stream of events

Beware the Culture of Excuses

Beware of the "red flags of poor leadership." Don't let these excuses go unanswered.

"What we do can't be measured."

"We are too busy to have that many meetings."

"We don't have the staff to collect all those numbers."

"We don't like to make our people uncomfortable in front of others."

"We are already doing that."

"We could find out, but it would take months, and we'd have to pull all our people off all their other work."

"We don't like to blame/embarrass our people here."

"We tried that, and it didn't work."

"That just wouldn't work here."

"We don't have the time/money to do that here."

"We only hold meetings when we need to here."

"It's different here."

(Beware any answer that ends with "here.")

issues from the decision, raising in one's favor all manner of unforeseen incidents and meetings and material assistance, which no man could have dreamt would have come his way. I learned a deep respect for one of Goethe's couplets: Whatever you can do or dream you can, begin it. Boldness has genius, power, and magic in it. Begin it now."

In chapter 7, I will lay out some straightforward steps for standing up your own performance management regimen—collecting data, benchmarking current performance, building out a permanent room for meetings, etc. But there is one foundational step required before you do any of those things, and it is this:

Set a date for the first meeting, and begin.

Recent history is littered with many well-intended mayors, governors, or other government executives who let their senior advisors and senior staff talk them out of beginning. Sometimes, the mayor becomes convinced that other priorities need attention first. The budget crisis (there is always a budget crisis), the pension crisis (there is always a pension crisis), the union negotiations (there is always the next round of union negotiations). It's not hard to convince an over-scheduled mayor that all these important things need to be tended to before a new performance management system can be implemented.

Other executives become convinced by their own staff that "we are already doing that," or (for whatever host of political, cultural, or equally dubious reasons) "it's different here," or "that just wouldn't work here." (See the sidebar, "Beware the Culture of Excuses.")

Still other executives allow themselves to become convinced that they don't have time "for that many

Collaborative leadership. Presence. Location. A working meeting in the midst of the Great Recession of 2009 with mayors from across Maryland—at their headquarters, not mine: the offices of the Maryland Municipal League.

meetings," when the truth is that a cadence of short, regular meetings focused on strategic goals is a time-saver. It eliminates the need for a whole raft of single-topic meetings, budget planning meetings, crisis management meetings, personnel meetings, and other meetings that masquerade as getting things done. And let's not forget how much time is wasted setting up those one-off, single-subject meetings, and corralling different people onto the same date and time on the calendar.

It's easy to be super-busy in government; getting important things done is hard.

In most governments, communication, coordination, and collaboration are unnatural acts between non-consenting adults.

Good leaders insist on repeatable routines: a rotation of short, regularly-scheduled meetings. It saves time. It drives performance. It makes effective collaborations not only more likely, but, as a practical matter, it makes them unavoidable. A cadence of short, regular meetings—with agendas, and after-action memos—focused on the big strategic goals is how big organizations achieve important goals. These are the practices and disciplines that hold everyone mutually accountable to one another for progress.

So, start and don't stop.

A relentless and repeatable pattern of short, regular meetings isn't about "increasing the frequency of the beatings." It's not about embarrassing or berating people. It's about setting a cadence of accountability for progress. It's about focusing on the trajectories for progress and measuring the daily path of that progress to the goal. It's about focusing on the leading collaborative actions that drive progress and deliver better results. It's about seeing what works.

And all this makes possible the second rule of this new way of leadership: **Lift up the leaders**.

Rule #2: Lift Up the Leaders

In any large organization, 10 percent of people are natural leaders, go-getters, and high achievers. On the opposite end of the bell curve is a very different group—the natural slackers.

That leaves 80 percent of us in the middle. Left to our own inclinations, the 80 percent of us in the middle will naturally lean back toward the behavior of the slackers. The great leadership challenge is to get the 80 percent in the middle to lean forward, not back; to lean forward in ways that follow the example of the leaders. The difference between an organization that rocks back and an organization that leans forward is the difference between nation-leading progress and business as usual.

How do you get the 80 percent to lean forward rather than back?

Good leaders do this by lifting up the leaders within the organization for all to see, and they do this by creating a culture of collaboration where hard work and achievement is recognized by all. The two are really just different ways of understanding the same essential practice.

Good leaders keep a compelling scoreboard from week to week so that everyone can see who is achieving the best results. This means making sure every team can see how well they are doing compared to all the other teams in a given endeavor, whether trash collection boroughs, building inspection zones, or police precincts. Put a map up on the wall so everyone can see it; shade it according to performance. Rank teams by objective criteria of performance.

Praise the leaders. Ask them how they did it. Tell them well done, and make sure everyone in the room hears the praise.

In our own work or team experiences, we've all known what it is like to be part of an organization or endeavor that totally lacks a collaborative culture. We describe these gray, lifeless situations as "not a fun place to work." Situational ignorance, mission-blur, blame, revenge, retribution, closed-mindedness, dishonesty, one-way communication, a lack of acknowledgment, and a refusal to listen: these are the human failings—made worse by poor or absent leadership—which undermine collaboration and teamwork.

But most of us have also seen and experienced—if sometimes only briefly—what it is like to work in a collaborative culture. To experience the joy of a shared sense of purpose, to receive acknowledgment of a job well done, and to feel the satisfaction that comes from working with others and "spending one's energies on work worth doing."

The Chinese have a saying: "Rivers and mountains may change; human nature, never." While the technology of measuring and mapping performance is a great enabler of creating a collaborative culture, the real driver is human nature. Collaborative cultures are brought into being and sustained by good leaders who understand how to harness human nature for higher achievement.

Most people work best against deadlines.

Jack Maple, the great New York Police Department savant of human nature and performance management, once tutored me on the underlying rationale for holding CompStat meetings in two-week intervals. "People work against deadlines," he said. "Do you want your organization to reduce crime one percent every year, or one percent every two weeks? Do you want people to work together to make the city one percent cleaner every year, or one percent cleaner every two weeks?"

A repeating pattern of short, regular meetings creates a cadence of accountability. It creates a steady work rhythm for achieving progress against last week's numbers. It creates a tempo of work against short, regularly recurring deadlines. (Refer to chapter 1 for more information.)

Small things done well make bigger things possible.

Most people like to be left alone at work.

Most people like to be left alone at work. I like to be left alone at work, too. But we call it "work" for a reason. Poets and artists get to work alone; the rest of us must work with others to be successful.

Unless the leader insists on it, communication, coordination, and cooperation don't just happen naturally across big organizations. Good leaders make these things happen. A good leader shows up herself in the center of the circle to make sure that it happens. And please understand when I say "circle," I am not only speaking metaphorically and operationally; I am speaking literally. There is a reason King Arthur had a round table—it worked. Meetings should be arranged for a conversation with people facing one another, not for a concert or lecture. When people are facing one another, they are far more likely to bring their "A game" to the problem-solving dialogue.

Given the shortness of time, the molehill of resources, and the mountain of complex challenges, there is no way to make progress in the face of tough challenges unless people work in thoughtful, collaborative ways with each other. One meeting every two weeks to focus just on public safety, or just on solid waste, or just on health, or just on housing is not a waste of time. It is essential time.

A repeating pattern of short, regular meetings creates the possibility for increasingly more effective collaborations. It allows tactics and strategies to be more quickly assessed and adjusted based on the latest emerging truth, the latest emerging evidence, and the most recent experience. It allows the best practices to emerge. And over the course of time, it lifts up the most effective tactics and strategies, and the most effective leaders.

Practical Advice to Future Leaders

The National Governors Association once asked me to share with an incoming group of Republican and Democratic governors the ten most practical pieces of wisdom that have served me best in governing. Almost all of them were collected from other servant-leader practitioners, most who were experienced executives. And all these things can be applied by leaders across the entire span of government.

1. The things that get measured are the things that get done.

2. Goals and deadlines are the x-axis and y-axis of all human endeavor.

3. Small things done well make bigger things possible.

4. A graph moving in the right direction is the most beautiful achievement in self-governance. The pace of progress is variable and it's negotiable, but the direction of progress is not.

5. Effective leaders make themselves vulnerable. Own the goals of the government you run and the people you lead—no one else will.

6. Whether a large human organization moves forward to achieve meaningful goals depends in large part on whether its leaders and achievers at every level are recognized by the chief executive and by their peers.

7. Timely, accurate information about performance and outputs *must be shared by all*—most importantly, it must be shared with the citizens you serve.

8. Communication, coordination, and collaboration are unnatural acts between non-consenting adults. Effective leaders create and enforce data-based routines of communication, coordination, and collaboration throughout their government. This is the cadence of accountability that only you can set.

9. "People make progress; common platforms make it possible." The geographic map of your community, city, or state is your common platform. All information systems must be based on the map.

10. We are not here to make excuses; we are here to make progress. Repeat this mantra over and over again—especially to yourself.

Most people do their best to honor commitments they make in front of a group.

It is important to always have a stenographer or note-taker at performance management meetings. It is important to produce follow-up memos itemizing commitments made during these regular meetings. And it is even more important to have your colleagues and co-workers witness those commitments.

It is much harder to ignore a request for help or information from a colleague that you are going to be sharing coffee with the next morning. For all the benefits of the internet and open data, there is no substitute for the integrity that comes from face-to-face dialogue and commitments openly made in the presence of all.

We are all busy, *and yet*, we all need to find better ways to operate as a team.

A repeating pattern of short, regular meetings—well-prepared and well-led—promotes a culture of mutual accountability. It promotes a culture of truthfulness and integrity in communications. And it promotes a culture of reciprocity and understanding within the group.

As a leader, after you start this process by setting the date for the first meeting, you can never allow the process to stop. As surely as the law of gravity—if you stop pushing the rock up the hill, it will most certainly roll back down. The hardest things to institutionalize in government are new systems that require constant work and tending. They are also the most important things to institutionalize when it comes to making lasting impacts and generational progress.

And this leads to the third rule of new leadership: **Lead with real-time awareness**.

*Preparing for hurricane landfall in the Situation Room of the State Emergency Operations Center. **Timely and accurate information shared by all is key to emergency operations management.** Short, regular meetings where responsible partners have the permission—and responsibility—to speak up and to ask questions throughout the emergency allow you to save lives by staying inside the turning radius of the event.*

Rule #3: Lead with Real-Time Awareness

I was once invited by the superintendent of our United States Naval Academy to address the entire brigade—all 4,000 of them—on the question of leadership, and specifically, the question of whether good leaders, *born or made*?

The answer, of course, is "Yes."

As it is with artists and athletes, good leaders are both born and made. All of us have some intrinsic leadership ability and we are all given different opportunities to hone those abilities over the course of our life and work experiences.

We might see the virtues of leadership—vision, future preference, courage, honesty, and integrity—as intrinsic to a person's character and upbringing. But the disciplines of leadership are things that can be taught and learned with experience. These disciplines include: the ability to articulate a vision, the willingness to take responsibility for the success of the mission, and the combination of courage, wisdom, and humility necessary to change tactics and strategies in pursuit of the vision.

In the Information Age, there is another essential discipline required for collaborative leadership, and that is to lead with a real-time awareness of the latest emerging truth. If there are no secrets anymore, why not embrace the new reality and make it work for your team?

The True Power of Leadership

Leaders can no longer sit high atop a pyramid of command and control where information is tightly held and hoarded. In fact, pretending that you can hold information from the public today is a fatal mistake. But, with a radical commitment to openness and transparency, leaders can put themselves firmly in the center of the latest emerging truth. Leaders can give themselves and their people the clearest, most up-to-the-minute, and most holistic picture of what is happening, and what is being done about it—real-time information about where, when, and why. It is the clarity and immediacy of knowing that allows the leader to focus their team and their collaborative circle—whether cabinet, division chiefs, command staff, or citizenry—on the latest emerging truth.

It is what battlefield and naval commanders call "situational awareness."

It is what New Age philosophers call "awareness."

And it is what citizens call "being on top of it."

The latest emerging truth is not a secret to be controlled but a power to be shared.

Effective leaders today must be more situationally aware than ever before, and they must be more present than ever before. Therefore, effective leaders put themselves routinely in the center of the collaborative circle and as close to the first outflow of the latest information—information about what is happening where and when. Effective leaders take up this position with total responsibility for what is. They hold this position even as they keep one eye focused on the next horizon. And all this means maintaining a commitment to openness and transparency that the old norms of politics would consider unwise or electorally risky.

This is the new positional advantage of effective leaders in the Information Age—knowing immediately what is happening where and when. Being among and with—at the center of the collaborative circle. And it is from this position closest to the latest emerging truth, that a good leader has the ability to exercise certain essential powers—powers that, even in the Information Age, are uniquely reserved to the leader.

Baltimore, Believe

In 2001, a commercial aired simultaneously on all the television stations in Baltimore. It was paid for by the private dollars of principled Baltimore business leaders. It began with the words of a little boy living life in a very hard-hit neighborhood in Baltimore.

"My grandmother says we're all part of one big fire. I don't know if that's true, but I know there's a fire inside me."

So began our very public campaign to awaken Baltimore's truer sense of self—to tap the fire inside—and to call upon the power of that spirit to confront the violence of drugs and drug addiction that was killing 300 to 350 of our young men—and, increasingly, our children—every year.

That jarring and disturbing commercial of the reality of drug violence and addiction in our city signaled the very public start of Baltimore's campaign—of Baltimore's fight—to "Believe."

You see, the day the campaign launched, Baltimore held the tragic distinction of being at or near the top of all the wrong lists: most violent, most addicted, and most rapidly abandoned major city in America. The beginning of that campaign was not a feel-good moment. There was nothing happy about that opening ad. The ad ended with that little boy's sister being gunned down on an innocent errand to the corner store, her young body lying lifeless in a pool of blood. I got a lot of calls from civic boosters and business leaders asking, "Why on earth did you run those?" But the ads that followed called upon all of us citizens to not only believe, but to act. Join the police department, get someone you love into drug treatment, mentor a child. The response was broad and deep.

From those days forward, Baltimore began to change for the better. Over the course of ten years, Baltimore achieved the biggest overall reduction of crime in any major city in America—bigger than New York or Los Angeles. In 2011, Baltimore, for the first time in more than three decades, reduced homicides to fewer than two hundred. Drug overdose deaths were driven down to all-time lows. Juvenile shootings were driven down 70 percent since 2007.

By 2012, Baltimore City public schools were posting their highest graduation rate—with an unprecedented 20-percentage-point gain in four years—and second-lowest dropout rate since we began keeping records.

In that same period, Baltimore's population decline slowed to a rate not seen since the 1950s. And notwithstanding the difficult recessionary years, Baltimore was rebuilding again, neighborhood by neighborhood, from the inside out.

These successes were not easily won. Too many Baltimore City police officers gave their lives for the hope of that safer city in which we called upon one another to believe. We never made perfect, and the mission was never totally accomplished.

But, together, we moved in the right direction. And to the cynical birds in the rafters who would like to dismiss Baltimore's achievements across three mayoral administrations as merely part of a national trend, think again. If you think smarter policing, better drug treatment options, youth interventions, and strong public funding don't matter, just look at other cities where crime is rising. (See chapters 4 and 5.)

Thinking back, it is hard to explain to young, new homeowners in growing neighborhoods like Canton, Bolton Hill, or Woodberry just how badly we had allowed apathy and acceptance of the status quo to destroy our belief in one another; how badly we had allowed our collective culture of cynicism to keep us from even trying. All the "smart" people knew that "it's just Baltimore—there's nothing you can do about it."

After years of shrugging our shoulders at the addiction and violence, our city came together in the Believe campaign to admit we had a problem; together, we started doing something about it.

Thanks to President Bill Clinton and Maryland's congressional delegation, we put two hundred more police officers on our streets. Thanks to the city council, we started paying them a lot better. Thanks to Governor Parris Glendening, we did what Mayor Kurt Schmoke had been urging for years and doubled funding for drug treatment. Religious leaders helped us recruit hundreds of volunteers to serve as mentors to city kids. And you know what? All that stuff, together, actually works.

Baltimore embraced the stark, white-and-black call to "Believe." The campaign took on a life of its own. Street vendors found people wanting to buy and wear Believe T-shirts. People placed bumper stickers on their cars. Believe trash cans rolled into once-forgotten neighborhoods. The Believe-mobile, sponsored by M&T Bank, toured the city and set up a lighted sound stage for neighborhood kids to play concerts while police closed the streets to the drug trade and opened them to the good people who lived there.

There was one big reason we ran that campaign: to challenge one another to believe—in ourselves, and in the fact that we are still the people whom Frederick Douglass and John Unitas loved. A people who believe that, together, we can make our city a safer place, a better place for kids to grow up.

The Power to Set a Shared Vision

Only the leader can set a shared vision. Others may inform the vision. Ideally the vision reflects the shared ambitions or needs of an entire people, but the leader must set the vision, own the vision, and make it clear to all that she or he will do whatever it takes within ethical bounds to pursue the vision. An important part of this is the willingness to set goals with deadlines.

In my administrations, I had many well-intended staff and advisors who begged me not to set goals with deadlines. Whether it was crime reduction or reducing pollution that flowed into the Chesapeake Bay, the traditional political wisdom holds that declaring a public goal with a deadline is politically risky to the point of being irresponsible. Concerned staff or cabinet secretaries would often ask, "What if we don't hit the goal?"

I would ask in return, "What if we do?" (This back-and-forth could go on for several rounds.)

Their concerns and fears were not only for their own jobs; they were also for mine. But there is only one way to know whether something is achievable, and that is to try. The difference between a dream and a goal is a deadline.

The Power to Convene

When it comes to pulling people together regularly in a collaborative pursuit of big goals, no one can make this happen but the leader. If you don't pull your team together, they will not do it on their own. And you might try to delegate the responsibility, but those meetings will soon cease to happen. In an age of information overload and multitasking carried to a level of compulsion, the power to convene is more essential than ever.

This one can be challenging. There are so many requests for time on a leader's calendar.

I have frequently heard it said by executives and their executive staffs, "We don't have time for that many meetings." To that time-worn excuse, I ask: "How much time do you have for meetings?" Sometimes the retort comes that "we only hold meetings when we have to." To which I ask, "How would you know when you have to?" Seriously? Do we wait until bad or tragic things happen? Do we wait until operational failings are so bad that they make it into the paper or blow up on social media?

When it comes to achieving the most important goals, no one else has the authority to convene the group except the leader. Just as no subordinate has the power to set the vision, no subordinate has the power to convene the team. This is not a function that can be delegated. The power to convene belongs to the leader.

The Power to Focus

As hard as most people work all week long, it is amazing how little time most organizations take to focus the collective experience and wisdom of the team on achieving the most important strategic goals. Only the leader can make this happen. Busy-ness is not effectiveness; don't be fooled. Effectiveness requires focus—the sort of focus that solves the puzzle. The sort of focus that deepens awareness and understanding. The sort of focus that encourages the team to ask more questions.

Good leaders constantly return focus to the most important goals of the mission.

As our own CitiStat process in Baltimore matured into its second year, I noticed a difference emerging between departments based on the leadership of the individual department heads. Most came to appreciate the value of a focused hour with the mayor's command staff to untangle impediments to progress and find better ways forward. But some department heads would come to CitiStat meetings as if it were a biweekly quiz rather than an active search for better ways to get things done.

When the questions about performance, cause and effect, or lack of improvement were asked, the red flags of poor leadership would fly—more excuses than answers.

"We don't have the staff to collect all those numbers."

"We could find out, but it would take months, and we'd have to pull all our people off all their other work."

But when Joe Kolodziejski, the head of the Bureau of Solid Waste, showed up to City Hall with his team for his biweekly CitiStat meeting, it was clear that his department had spent some time, effort, and imagination since the last meeting focused on overcoming their big challenges. It was clear that operationally, the leader of Solid Waste pulled his team together on a regular, daily basis to focus on their important strategic goals; to ask more questions; to work the problem; to think.

Good leaders use their power to focus the team's imagination on achieving goals.

The Power of Presence

There is no substitute for the presence of the leader in the center of the collaborative circle. Being among and with is not a luxury, it is a necessity. This is not to say the leader is able to make every single meeting. But she or he attends enough of them, so their presence is always felt. And when she is physically present, the leader is so entirely awake, aware, responsible, prepared, and "here, in the moment" that everyone in attendance understands intuitively the importance of the focus, the importance of the mission.

As mayor and as governor, I quickly learned there was no substitute for my own personal presence—even if only for a portion of a meeting. To praise a job well-done. To return focus to the larger goal. To admonish or push when communication and collaboration seemed to break down. To spur the team forward when the rough patches were hit.

But I also learned there were surrogates for those times when I was not able to be 100 percent present. One was to have a chief of staff or chief of operations who was always a driving force in center of the circle. Making all understand the trust and the constant, timely flow of communication between the two of us was presence by proxy. Thank you notes for a job well-done was another way of being present. Writing notes with follow-up questions in the margins of the executive briefing memoranda that accompanied any CitiStat agenda was another way to show presence. Showing up for a few minutes to observe a short portion of the meeting, if only from the back of the CitiStat room before rushing out of the building for another event, was another way.

Being aware of progress to goals and being aware of barriers to achievement. Understanding the changing dynamic of a problem, whether it's a rash of burglaries, a weather event, or a spike in overdose deaths. Knowing who the high achievers are and lifting them in front of others.

These are all ways a good leader exercises the power of presence.

The Power of Belief

If you believe you can or you believe you can't, you are probably right. This is true of individuals, and it is true of organizations that take their cues from the leader. Tell your team you trust them, you are proud of them, you need them, and you are counting on them. Remind your team of the smaller accomplishments that make the next big thing possible. And before and after all those things, tell them you believe in them and you believe that together you will succeed.

One of the easiest beatdowns in public service is the tyranny of low expectations. How many times have we heard people snicker or mutter phrases like "close enough for government work?"

Using Your Eight Drivers to Overcome Today's Challenges

Presence is really a baseline—or foundational—energy, one that drives all traits and functions necessary to live your best life. It also drives the other seven drivers. When you live in the present moment, you understand that everything is connected. Irrelevance is gone. Everything matters. You absorb every bit of life because you are highly focused. You think more clearly and efficiently. You act with more integrity and clarity. You are unburdened by unproductive thoughts of the past or future. You worry less. You fear less. You are infinitely more creative.

Openness is the second key driver of your best life. Many, if not most of us, have learned through difficult life experiences to resist "what is." But resisting "what is" causes more pain and drains our energy. Opening to "what is" becomes liberating and energizing. When you're open, you constantly seek to widen the net for possibilities, and resist nothing. In your best life, you are curious, you are a font of ideas and creativity, and you see possibilities everywhere.

Clarity is the third driver of your best life—clarity in thoughts, emotions, and behavior. We have all, at least on occasion, thought, emoted, or acted out of anger, rage, envy, insecurity, guilt, greed, or some other fear-based stimulus. But when you are clear, you find it easy to define every element of who you are, both to yourself and to others. You are people oriented, open-hearted with a genuine love for people. You see the good and the potential in everyone, instead of a threat. You have healthy, empowering relationships with others.

Intention is the fourth driver of your best life. In every moment, each of us can choose intention or neglect, intention or disempowerment. While many of us constantly say or think "I hope" and "I want" and "I'd like," few of us sincerely believe we can bring about a desired result. Practicing intention, which involves a discipline of expressing your desired result in great detail, regularly visualizing it as a current reality, offering exchange for it, starting a "conspiracy" of people focused on helping you achieve your intention, and, ultimately, detaching from it, significantly helps you to achieve the results you want in your life.

Personal responsibility is the fifth driver of your best life. We live in an era where personal responsibility has been replaced by blame and litigation. These are fear-based responses. Personal responsibility is complete ownership of "what is," as distinguished from openness, which is the unbounded willingness to consider every element of "what is." When you learn to own "what is" on every front and create the energy that results when you can say, "I am completely responsible for every positive and negative element that exists in my life," you will see a dramatic improvement in the integrity with which people view you, your courage, and your personal relationships.

Intuition is the sixth driver of your best life. Each of us was gifted with a powerful source of inspiration—a knowing, an intuition—that is embedded in this omniscient energy that binds everything that is. But fear often causes us to abandon it too quickly in favor of a "safer" route supported by "facts" or the opinions of others. In doing this, we abdicate the crucial role that active intuition plays in life. The skilled and liberal use of intuition enables your ability to make good decisions in all areas of your life, adapt to uncertainty and changing conditions, and interact with others in a highly empathic, supportive way.

Creativity is the seventh driver of your best life. If you want your best life, when you truly appreciate that life is binary—there is only creation and destruction, growth and decay, life and death—it is pretty easy to decide that you want to be on the side of creation, growth, and life. The key then, is stoking your creativity in every possible way so that you remain aligned, and not at odds, with life itself. Fortunately, every person has the potential to be a powerful creative force. When you tap into that creativity, you become highly energetic, you see possibilities instead of barriers, you see a better life for yourself and everyone around you, and you see a path for achieving it.

Connected communication is the eighth driver of your best life. In the complex, adaptive system in which we live, where everyone is interconnected, and relationships are paramount, communication is essential for survival. Better communication is a function of increasing the connection in your communication. "Connected communication" is an intensely powerful energy—a driver—deep within each of us. The system of connected communication, from clear expression of a purposeful message by an empathic speaker to an empathic listener, fuels your ability to be supportive of and inspiring to others and have productive, empowering personal relationships.

Conclusion

The solution to our worsening societal woes, and conversely our best societal life, is in leadership. Leadership at the societal level begins with leadership of the self. Each of us living our best individual life will flow into our best collective life. It is just a matter of accessing powers—drivers—with which we are already endowed.

—David M. Traversi, author of *The Source of Leadership*

But you will never find an effective leader in any walk of life or in any human endeavor who accomplishes anything difficult and meaningful—whether it is a Super Bowl victory, a winning election campaign, or making a city safer, cleaner, and healthier—without first instilling in his team a belief that they can win.

Belief of the leader is a powerful force in the group dynamic. There are tools that can amplify that belief, but there is nothing that can substitute for it.

The Art of Collaborative Leadership

One of the quirky side benefits of serving as a governor or mayor is that at least once a month a new book on "leadership" lands on your desk—a complimentary copy sent by a publisher, author, or promoter.

I found most of them useless.

But the one I use in my courses is not. It is *The Source of Leadership: Eight Drivers of the High-Impact Leader*, by David M. Traversi. It is practical, real, and good—a description which also applies to the young Americans I've had the honor to teach at the University of Maryland, Boston College Law School, Georgetown University, and Harvard University.

Traversi's "eight drivers of leadership" are *presence*, *openness*, *clarity*, *intention*, *personal responsibility*, *intuition*, *creativity*, and *connected communication*. All these drivers are important, but "presence" is the driver which makes all the other drivers possible. (See the sidebar "Using Your Eight Drivers to Overcome Today's Challenges.")

Knowing what these qualities are is one thing. Developing them within yourself and others is another. Calling them forward in a big organization is yet another. These are the soft skills. The skills to be practiced and improved upon with every important and passing day. They are the skills that come together in a mix and balance unique to every leader.

In a brilliantly insightful address to US Army cadets at West Point in 2009, Professor William Deresiewicz said on the subject of "solitude and leadership":

> "... the great books, the ones you find on a syllabus, the ones people have continued to read, don't reflect the conventional wisdom of their day. They say things that have the permanent power to disrupt our habits of thought. They were revolutionary in their own time, and they are still revolutionary today. And when I say 'revolutionary,' I am deliberately evoking the American Revolution, because it was a result of precisely this kind of independent thinking.
>
> "Without solitude—the solitude of Adams and Jefferson and Hamilton and Madison and Thomas Paine—there would be no America."

And so it remains with the practice of leadership today. The eight drivers are important, but doses of solitude and deep reflection are essential too.

So start and don't stop.

Lift up the leaders.

Lead with real-time awareness and from the powerful silence of your own heart.

Learn & Explore

Leadership Forum
Watch a video of my presentation at the Frank Batten School of Leadership and Public Policy in 2017.

NewDEAL Leaders
The NewDEAL (Developing Exceptional American Leaders) is a national network of state and local leaders working to expand opportunity for all Americans in the changing economy.

Baltimore Believe TV Commercial
Watch the original 2001 video that signaled the very public start of Baltimore's campaign—of Baltimore's fight—to "Believe."

For links to these and other examples, exercises, and resources, visit SmarterGovernment.com and click chapter 3.

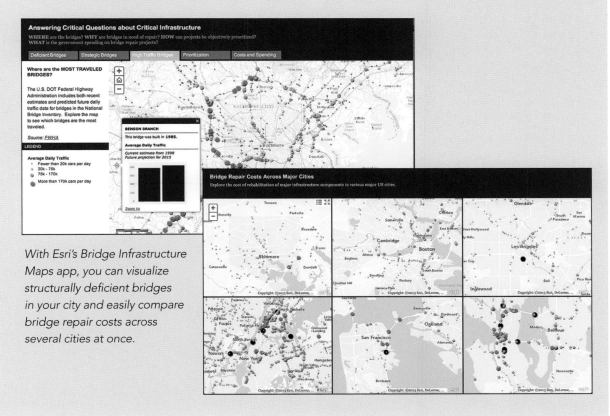

With Esri's Bridge Infrastructure Maps app, you can visualize structurally deficient bridges in your city and easily compare bridge repair costs across several cities at once.

4

A New Way of Policing

The Information Age has given us many new technologies that have in turn brought us new opportunities to change the way we govern our cities, states, and countries. This is perhaps best exemplified by our ability to measure and manage all elements of performance statistically and geographically, a process pioneered by Jack Maple in the New York Police Department.

A Canon for Crime-Free Communities

CompStat was the beginning of a new way of policing in the United States; it was also the beginning of a new way of governing.

Pioneered by Jack Maple of the New York Police Department under the leadership of NYPD Commissioner Bill Bratton, CompStat harnessed the power of GIS for real-time intelligence and situational awareness shared by all.

Wherever crime was happening, the dots on the map told police commanders how to deploy to get in front of it. The repeatable routine of collaborative, questioning meetings—focused on the map—helped commanders figure out how to prevent crime and how to solve crime.

Reducing Crime in the Information Age

One of the great but little-acknowledged success stories of the last thirty years is the record reductions that have been achieved in violent crime. There are far fewer victims of violent crime today in the United States than there were fifteen years ago. In city after city, collaborative police leadership combined with new technologies like GIS are saving a lot of lives.

We have a bad habit as a country of always blaming our police when crime "goes" up, but never giving our police credit when crime "goes" down. Even the language we use to describe crime trends is full of passive verbs that are more appropriate for describing weather events than human events. We generally say that "crime is up," or "crime is down" as if we are describing trends in temperature, rain, or barometric pressure.

Crime doesn't "go" where it likes as if it were the wind. Violent crime is driven up by the intentional acts of human beings who choose to rob, choose to shoot, choose to steal, or choose to otherwise harm their neighbors. Crime is driven down by a countervailing host of other intentional actions, none more essential and foundational than the operations of a well-led police department. Among those actions are solving crimes, apprehending offenders, and successfully removing repeat violent offenders from the streets.

Yes, there are many things that each of us can do to deter and prevent crime. And there are things we must do as a people to reduce the root causes of hopelessness that contribute to crime. Improving education is essential. Creating jobs and economic opportunity is also essential. There are many actions that build a just society. But foundational to all those is the need for safety and security.

Philosophers and social scientists may theorize all they like about the degree to which socio-environmental issues—employment, cell phones, the removal of lead from gasoline, legalized abortion—are responsible for record reductions of violent crime in the United States over the last three decades. There is still much work that remains to be done to reduce incarceration rates, recidivism, and the use of lethal force. And clearly, we must do more to reduce drug addiction and improve re-entry programs. But none of these things can really be very effective unless the police are doing their part to protect our lives and safety.

From 2000 to 2009, three major cities led all other cities in America in the rate of reduction of violent crime. Those cities were: Los Angeles, New York, and Baltimore. And the common thread that ran through those record reductions was CompStat—a new and better way of policing.

The great promise of effective governance in the Information Age is not so much that data allows us to manage the masses but rather that it allows us—if we care—to keep sight of the security and dignity of every individual person.

Measuring performance.
Getting things done.
Bringing people together.
A new way of governing.
And it all started with CompStat.

The Rise of CompStat

I was on the Baltimore City Council when I first heard about CompStat and how well New York City was using it to dramatically cut crime. Soon, I was reading everything I could find about Jack Maple and the new system of crime-fighting he had first invented as a transit police lieutenant policing a portion of the massive subway system of New York City.

New York City Police Commissioner Bill Bratton recognized Maple's talent and insights early on. He plucked him from the subway system and made him deputy commissioner in charge of crime control strategies for the NYPD. It was very likely the biggest single promotion in all of American policing. It was

What Is CompStat?

CompStat is an entrepreneurial, collaborative, and iterative process of performance-managed policing. Initially called Charts of the Future and using stick pins on paper maps, CompStat (for COMPare STATistics, the name of the original computer program that automated the manual process) is a methodology for analyzing patterns to fight crime in New York City. It was pioneered by the late Jack Maple.

The four tenets of CompStat are:

• Timely, accurate information shared by all,
• Rapid deployment of resources,
• Effective tactics and strategies, and
• Relentless follow-up and assessment.

A key component of CompStat is the use of GIS as a common operating platform to focus the best thinking of the organization on emerging crime trends in real time.

In addition to computerized pin mapping, a key feature of CompStat is weekly accountability sessions, where personnel from every precinct present summaries of crime statistics and maps to agency executives. This allows the executives to review performance, set strategy, and hold individual agencies accountable for good—or bad—results.

CompStat helped the NYPD dramatically reduce crime and is employed today by several police departments around the world.

from this leadership post that Maple pioneered the NYPD's CompStat program—with nation-leading results. But no police commissioner lasts forever. When Bratton left in 1996, so did Jack Maple.

Jack became a consultant—a crime reduction expert, as his matter-of-fact business card stated. From Baltimore, I remember reading about Jack's arrival as a police consultant in New Orleans, where he would orchestrate a short-lived but equally dramatic turnaround. The Big Easy was not easy for Jack. But neither was New York—or Newark, New Jersey, where Jack had also done some work. Tough as they were, these were places with the greatest potential for saving lives. These were, therefore, the places to which Jack was drawn.

"I'd really like to meet this guy one day," I thought.

Who Was Jack Maple?

Once upon a time, in the days before MapQuest and Uber, there lived a 5-foot-7-inch, slightly round lieutenant of the New York City Transit Police who created what he called "Charts of the Future." That time was 1991.

With paper and color markers, the lieutenant would plot where and when robberies took place on his part of the subway system. Then he would deploy undercover detectives and transit police officers to catch the robbers where and when the maps showed they were striking.

Soon, the lieutenant was driving down robberies to record lows in his part of the subway system. His success and his "Charts of the Future" earned him the cover of the Sunday *New York Times Magazine*. His talent also caught the eye of the new police commissioner of the NYPD. In 1994, this New York City Transit lieutenant was made deputy police commissioner for crime control strategies for the entire NYPD.

He replaced his colored markers, paper, and acetate with Esri's ArcGIS technology, and suddenly, CompStat—and a new way of performance-measured policing—was born.

That man of imagination and insight was Jack Maple.

And the NYPD, under his and Commissioner Bill Bratton's leadership, went on to reduce violent crime to levels that few would ever have thought possible just a few years before. New York's ongoing success in reducing crime and saving lives, led to a revolution of performance-measured policing in cities and towns across the United States.

After retiring from the NYPD, Jack Maple became a "crime reduction specialist." He helped cities across the United States and around the world replicate CompStat. One of those cities was Baltimore.

With Jack Maple (right) at the John F. Kennedy School of Government at Harvard University in 2001.

The CompStat approach was empowered by advances in computational technology. Instead of waiting for a cop on light duty to punch colored pins into a map mounted on cork board to figure out where crimes were happening—long after the fact—simple off-the-shelf software and a computer could now do it instantaneously. GIS technology could now give police a complete, real-time picture of where crimes were being committed very close to real-time. This allowed departments to deploy officers immediately to where crimes were happening. And all the different divisions of the police department—patrol, detectives, vice, narcotics—could instantly share the information on a map—a GIS map—making it easier to identify trends, connect the dots of a case, and lock up the bad guys.

At home in Baltimore, where we were watching open-air drug markets expand, homicides rise, and citizens flee, a few of us on the City Council started asking: What can we learn from New York? What are they doing that maybe we should try?

In 1996, I was tasked by the Baltimore City Council president to lead a City Council delegation to New York to see CompStat first hand and come home with our findings.

The Four Tenets of CompStat

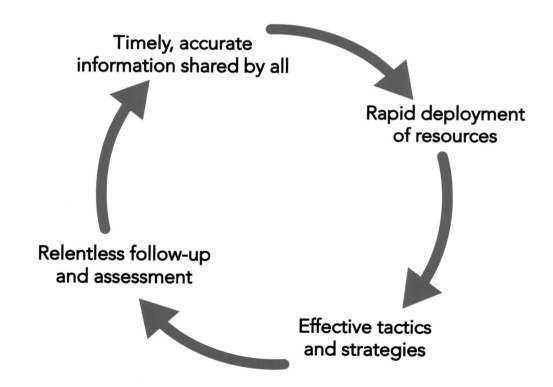

Governing for Results: CompStat in New York City

With the introduction of CompStat in 1994, New York City police were deployed based on the latest patterns of criminal activity with the goal of anticipating and preventing crime.

From 1993 to 1998, murders in the city were reduced by 67 percent, and robberies were reduced by 54 percent, well ahead of national averages and down to a rate not seen since the 1960s.

CompStat is still in use today, more than twenty years later—and still showing results. In 1993, the year before CompStat first came into use, the city had 1,946 murders. By 2015, it had just 352.

NYPD's Times Square substation enhances the ability of the department to connect with New Yorkers and protect the millions of people who visit Times Square annually.

A Culture of Accountability

A true CompStat meeting in action is not a sleepy weekly report; it's an intense, focused, hour-and-a-half long search for solutions. And there is a single source of truth at the center of it all: the numbers of crimes displayed on the map. Crime maps create a picture. Crime maps create a laser focus on the critical issue of reducing harm to citizens.

At the meetings our council delegation witnessed that day in 1996 at police headquarters in Manhattan, every commander owned his numbers and the reality they portrayed. I mean this both literally and figuratively—each commander stood at the podium in front of their numbers, which were projected onto the big screen behind them. And there was a picture of the commander on the screen in case anyone wondered whose responsibility those numbers were.

It was not so much a grilling as it was a collaborative dialogue. It was sometimes tough, sometimes pointed, always demanding, and always a call for collaborative leadership at every level of the department.

The deputy commissioner—and not infrequently, the police commissioner himself—sat in the center of the command staff horseshoe. On his right was the chief of patrol for the entire city. The two of them drove the questioning back and forth as they progressed through the agenda.

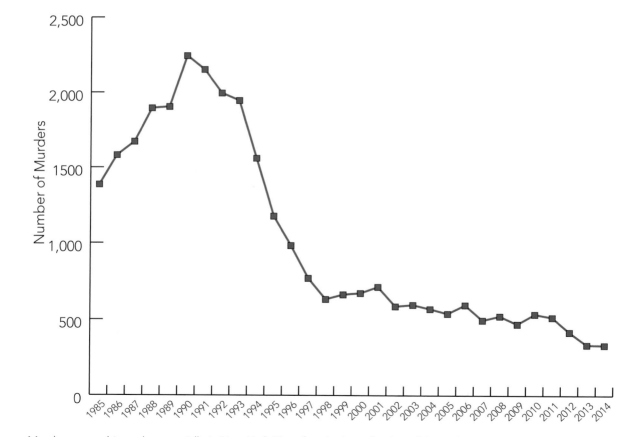

Murders were driven down rapidly in New York City after the introduction of CompStat in 1994.

Jack Maple on Leadership

At an Indian restaurant on Lombard Street in downtown Baltimore over a working lunch, Jack Maple shared his leadership wisdom with me. And for fifteen years as mayor and then governor, I put it into practice.

"One of the leader's most important functions," Jack said, "is to lift up the other leaders across the organization. Why? So everyone in the organization can see who they are...and why is that important? For one very good reason: it's called human nature."

Jack then put his elbows on the table and held his forearms straight up and down like a goal post as he segmented an imaginary bell curve of human behavior into three parts.

"In any big organization of people, 10 percent of us are inclined to be leaders, 10 percent of us are inclined to be slackers, and 80 percent of us are somewhere in the middle. It's like a bell curve. The 80 percent of us in the middle could lean forward toward the leaders, or we could kick back toward the slackers. If nobody at the top cares, if no one is measuring or keeping score, the 80 percent of us in the middle will naturally lean back—and slacking becomes normal."

With this, Jack tilted his arms framing the 80 percent backward—toward the slackers. "But if the leader at the top actually measures performance...if he or she brings people together relentlessly to point at the scoreboard, to see the names and the faces of the people who are achieving not just good, but better than good, then those of us in the middle will tilt toward the leaders."

And with this explanation, he tilted his arms framing the 80 percent in the middle forward—toward the 10 percent who are leaders. "The difference between the backward tilt and forward tilt is the difference between business as usual and nation-leading progress.

"People are dying here because of business as usual...so, measure performance across the board. Make sure everyone can see who the leaders are and why. Pat them on the back. Praise them in front of their colleagues and peers and citizens. Send them a 'thank you' note every now and again for a job well done. Make sure everyone knows they got the note. And promote the leaders every chance you have.

"Don't ever forget, as mayor, one of your most important jobs is to lift up the leaders so everyone can see who they are and follow their example."

"It looks like you've gotten on top of the rash of home burglaries in the southern end of your precinct since last time we met, commander. But I see another cluster of burglaries emerging just north of those burglaries. Did you apprehend and charge a suspect, or just shift it? What time of day are most burglaries happening? Is the perpetrator using the same or different methods of entry? What is being taken? Have you checked the pawn shops? Have you alerted the neighborhood watch? If you think it is kids, have you checked the truancy rolls at the nearest high school and middle school? Let's move on to robberies..."

Also seated at that horseshoe was the chief of detectives, the head of Legal Matters, the head of Internal Affairs, the head of the Fugitive Task Force, the chief of Transit, the head of Human Resources, and the head of Budget Management. Thus, if a problem came up that required one or more members of the city command staff to work it out, they were already all right there. There was no confusion and no delay. And there were no meetings three days later to talk about what might have been said at the meeting three days before. A stenographer took detailed notes, and each participant got a memo enumerating follow-up items to be resolved by the next meeting.

This was CompStat.

As our council delegation looked on as silent observers, we saw that CompStat was different. It was timely and accurate. It was open and transparent. It was real time. And it was unlike anything we had ever seen in Baltimore policing.

Just a few short years before, many people referred to New York City as the "Rotten Apple" instead of the "Big Apple." There were Hollywood movies made like *Escape from New York*. The plot in that movie was that ever-escalating violent crime finally forced our federal government to just build a big wall around New York City to keep the out-of-control crime inside.

But in the New York of the late 1990s, CompStat was saving hundreds of lives. It was giving all the citizens of New York—rich and poor—safer neighborhoods and a safer shared future. It was gaining traction and being replicated in other places. It was the method of performance management that would one day totally change for the better the way we would run our own City of Baltimore. It is the system that, in some name or form, has now been adopted in most large cities in America—and in hundreds of smaller and mid-sized cities around the world.

Collaborative Leadership in Action

With CompStat, everyone knew the goals. Everyone could see the scoreboard. Daily action was the expectation. And the range of possible actions—with everyone involved bringing their intelligence and capabilities to bear—was almost limitless. The cycle of regular meetings created a cadence of accountability. It also created a meritocracy as the most effective leaders over time were promoted through the ranks based on their performance. And the openness of the process meant that everyone knew who the most talented leaders were; most wanted to follow and emulate the leaders.

On deck the day of our council delegation visit from Baltimore, one precinct commander and his own precinct command staff were front and center with projector screens of their neighborhood maps and crime numbers behind them. In the wings, another precinct commander would observe his colleagues at bat. We learned afterward that the NYPD management tried to pair a more experienced, higher-performing precinct commander with a less experienced or struggling commander on the same day so that the second group could learn by listening to the CompStat dialogue of the first to see what was working and what was not.

CompStat is an iterative process. It is an ongoing, questioning dialogue held in regularly recurring

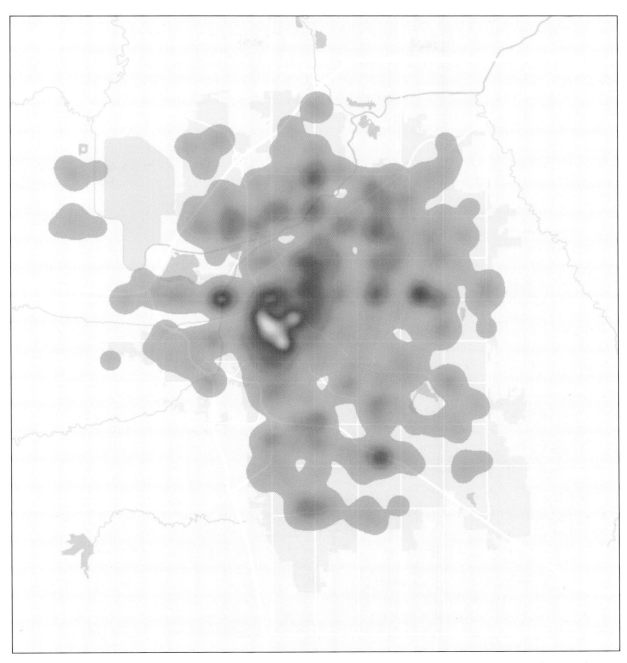

Since geospatial science entered the law enforcement arena, America's police departments have used GIS tools to direct patrols toward crime hot spots. Known as putting "cops on dots," this predictive-policing strategy has reduced some crime rates by allowing departments to concentrate officers on troublesome areas instead of randomly patrolling. In this map, hot-spot neighborhoods are easily identified so police can more efficiently tackle lawbreaking.

intervals between emerging realities and the tactics brought forward to address them. The iterative process constantly asks the fundamental question: Is what we are doing working? Or is it not?

For this reason, CompStat is fundamentally entrepreneurial, not ideological. It is an ongoing, longitudinal experiment. If things work, the evidence of success would suggest that we should do more of it. If things are not working or bringing about other unintended consequences, the evidence would suggest we should stop doing it, adjust, and try something that might be more effective.

Is it working to save lives and reduce crime? Or is it not?

That is the daily operational question of CompStat.

As I listened to the conversation—watching police accountability in action—I couldn't help but wonder: "Why doesn't every police department operate this way?" And more importantly, "Why doesn't every department of every government operate this way?"

From Pushpins to Computer Pins

If you have ever watched an old police movie or detective show, you might have noticed the maps on the walls. Often, these maps were mounted on Styrofoam or cork, so one could put color pushpins into them. Each pushpin represented a crime that happened at that place. In the nonfictional version of these maps, an enterprising person could get creative and have different colors for every major crime. And maybe even a different map for every shift.

The problem was that maintaining a map like this before the Information Age required a very enterprising and diligent person indeed. Crime reports were filled out by pen on paper. The reports had to be collected. The locations had to be looked up. The pins had to be placed manually, one at a time. Everything about the process was laborious, and nothing about it was real-time.

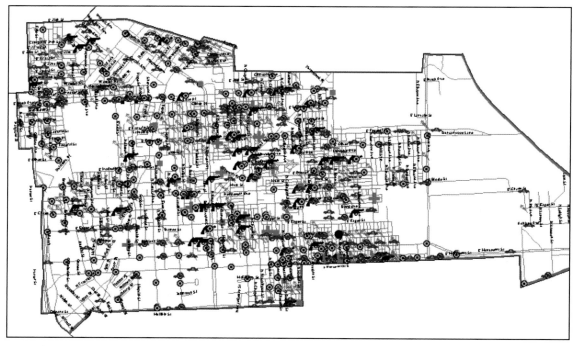

An early CompStat map from Baltimore based on NYPD CompStat iconography showing crime incidents across the Eastern district of Baltimore on a year-to-date basis.

And who was responsible for making the map? Instead of an enterprising and diligent officer, it was usually a duty relegated to whatever officer was on light duty with a sprained ankle or knee (or, often, a bad attitude) that had to be taken off the street for a while.

One can only imagine the time lag between when a person was victimized and when a pin representing that specific time was placed on the map. Let alone the time lag between when a series of rapes take place and when—if ever—pins on the map allow a questioning mind to see an emerging pattern of time or place.

The Information Age changed all that.

Simple, affordable off-the-shelf software like Esri ArcGIS and the basic computational technologies of Microsoft® Excel® spreadsheets changed push-pin mapping to computer-pin mapping. The speed went from "someday, sometime," to real time, all the time.

This new map, overlaid with real-time data from the 911 dispatch and call centers, has now become the common operating platform for modern police departments everywhere. Thanks to new technologies like GIS, the internet, and basic statistical computing, the emerging truth can be made very clear and visible for all to see.

The emerging truth for a police department is where and when crime is happening. If a police department is well-led, being aware of that emerging truth and getting on top of it—and ideally, ahead of it—is the responsibility of every commander and every member of the force.

Understanding precedes action. And better and more timely understanding precedes better and more timely actions.

Every conversation in every CompStat meeting is a mixing and mingling of the four primary tenets first laid out by Jack Maple:

1. Timely, accurate information shared by all
2. Rapid deployment of resources
3. Effective tactics and strategies
4. Relentless follow-up and assessment

These are not proverbs. They are maxims, they are disciplines, and they are the evidentiary basis for answering the most fundamental questions:

"Do we know, and are we doing something about it?"

"Are we reducing crime, or are more of our citizens being harmed?"

"Is the tactic or strategy working, or is it not?"

"Beware," Jack Maple would later admonish me, "the phrase which most people want leave out of the equation is the one about information being *'shared by all'*… Unless everybody knows, you can't get everybody to work. This stuff requires a radical and relentless commitment to openness and transparency—for the police and the public alike."

How CompStat Changed Policing for the Better

CompStat changed policing for the better in several significant ways.

1. By sharing timely, accurate information with all, CompStat advanced the development of collaborative problem-solving as a routine function in modern policing.
2. By enabling a more rapid deployment of resources, CompStat made it possible for police to better anticipate and prevent crime. It allowed police to get inside and disrupt the arc of emerging crime patterns.

Interactive Crime Dashboards for

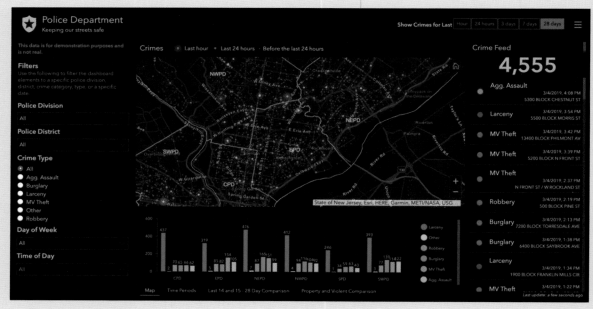

Building on the innovation and legacy of CompStat, state-of-the-art dashboard apps today provide a variety of ways

Police and Citizens

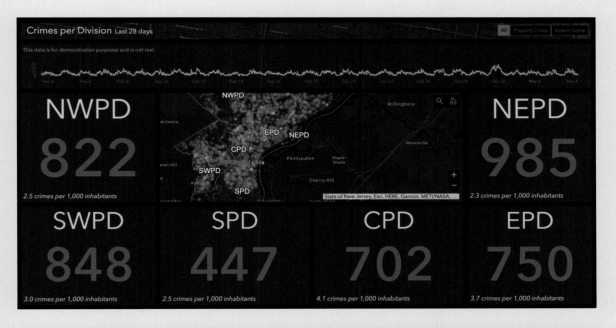

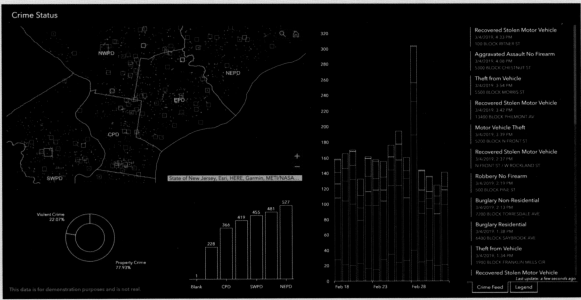

for police and citizens to interact with crime data and visualize crime patterns in their communities.

3. By convening the collective experiential wisdom of commanders together in short regular meetings to discuss the changing dynamics of crime, CompStat marshalled the best available thinking of the group. It tapped the collective experience of leaders to evaluate and share effective tactics and strategies.

4. By increasing the operational tempo for delivering more timely results. Rather than measuring results against annual deadlines, the entire NYPD started to work against two-week deadlines. Through this relentless, ongoing cadence of operational meetings and two-week deadlines, leaders kept the organization focused on the primary mission of improving public safety.

5. By enforcing repeatable, collaborative routines to focus on what works. As CompStat measured and compared the lagging indicators of public safety—namely reported crimes—in two-week increments, it also created an ongoing, regularly predictable rotation of command staff and precinct-by-precinct leadership meetings. The collaborative focus on leading actions—measured every two weeks—accelerated the tempo of operational progress to goal.

6. By fostering a more innovative culture. The CompStat methodology is both entrepreneurial and agile. It is entrepreneurial because the dialogue at every meeting is focused on what works. It is agile—in the innovative sense of that word—because with the short intervals between meetings and the immediacy of feedback loops, the collaborative circle of commanders was able to try new approaches and see quickly whether they delivered better results.

7. By giving NYPD leadership the performance data they needed to build up a meritocracy within the NYPD and the ability to lift up leaders throughout the organization for all to see based on actual performance. This not only improved the quality of supervisors and the likelihood of meritorious promotions, but it also improved department morale and effectiveness.

As a young Baltimore City Council member in 1996, I watched that CompStat meeting in New York unfold with a strange combination of emotions. My heart and my mind sparked with hope, optimism, and curiosity about this new art of the possible. And yet, the leadership and management disciplines we saw on display before us were things that seemed so far beyond the grasp of the current politics of Baltimore.

When we returned home, we published our report. Held a hearing or two. Drove some articles and some letters to the editor into the local paper. And then, we went back to policing the way we had always done it. Every week on the council floor, I'd remind members and those watching the city council cable channel that we did not need to accept the notion that there was nothing we could do about violent crime. I gave it my passionate best; but, most nights it felt as if I might as well be fighting the waves.

Just when you think things will never change, they change.

And sometimes, they change for the better.

Learn & Explore

CompStat: An Inside Look at the NYPD's Crime-Fighting Tool
A segment from NBC Nightly News about CompStat: "the crime-fighting tool that helped turn New York into America's safest big city, requires police officers to question not just suspects but each other."

CompStat in Practice: An In-Depth Analysis of Three Cities
A report from the Police Foundation analyzing the experiences of Lowell, Massachusetts; Minneapolis, Minnesota; and Newark, New Jersey, using CompStat.

CompStat: Its Origins, Evolution, and Future in Law Enforcement Agencies
A report from the Bureau of Justice Assistance and the Police Executive Research Forum.

For links to these and other examples, exercises, and resources, visit SmarterGovernment.com and click chapter 4.

Learn how to analyze robberies in preparation for a CompStat meeting.

5

Making Baltimore Safer

By 1999, Baltimore had become the most violent, addicted, and abandoned city in America. But by the mayoral race later that year, we were all sick and tired of open-air drug markets and violent crime. We were looking for new leadership to pull us together, and that is what we voted for. We were looking for new ways to make our city a safer place.

And CompStat delivered.

Campaigning for Crime Reduction

On June 20, 1999, at the age of thirty-six, I announced that I was running for mayor. With my wife Katie and just a handful of friends at my side, my announcement speech was short and to the point.

"...Hear me, Baltimore. Six months after I take office, the open-air drug market of this corner and nine others will be things of our city's past. In the second year, twenty more open-air drug corners will likewise be shut down, and, thus, will the people of this city easily measure our success or failure. When we make fighting crime and closing down open-air markets the top priority of Baltimore City government, then, and only then will we be able to build a stable and growing city tax base. Then, and only then, will we dramatically improve schools. Then, and only then, will the new jobs created by increased private investment be things of our city's present and future..."

The day I announced, I was the first choice of just 7 percent of my neighbors. My two primary opponents were both African American office-holders in a majority African American city, and both polled with more than 80 percent name recognition.

There would only be eighty-six days from announcement day to the primary election. Every single day of the campaign was difficult. Every conversation of the campaign involved difficult questions of racial injustice and law enforcement. Citizens rightly demanded to know how I intended to hold the police accountable to the law and the Constitution as I pushed them at the same time to shut down open-air drug markets. The term "zero-tolerance" became a double-edged term. For some, it meant we would no

In the thirty years leading up to 1999, Baltimore lost more of its population than any other major city in America. Across whole neighborhoods of East and West Baltimore, open-air drug markets and violent crime accelerated the flight—leaving behind thousands of vacant and abandoned homes.

longer tolerate open-air drug markets to terrorize our poorest neighborhoods. For others, it meant we would run roughshod over individual freedoms and constitutional rights.

We released two policy booklets in the early weeks of the campaign. The first was exclusively focused on how to reduce violent crime and hold the police accountable. The second was about everything else—improving public schools, job creation, housing, public health, parks, and transportation.

At an early candidate forum in front of Baltimore's business leaders, the three of us were asked if any of us would commit to cutting crime in half during our term of office. Only I said yes. The others had many reasons why no one could really make a promise like that. I answered with a plan. I also asked that they hold me accountable to that commitment.

We started to pick up key endorsements of black and white neighborhood leaders across the city, and then even some elected officials started to endorse. The campaign grew every day. It grew to fill an already large consensus. For most of us had come to believe that until we made our city safer, our city didn't have much of a future.

In the final days of the campaign, one of my opponents sent out a jumbo postcard mailer with a grainy picture of Rodney King being beaten by white police officers in California. It asked in large print: "Are you ready for zero-tolerance?" On the back of the postcard was a picture of my white face.

The *Baltimore Sun* endorsed one of my opponents. Police and fire unions endorsed the other. But on Election Day, we won with 54 percent of the vote. In fact, we won each of the six City Council districts of the city—defeating both opponents in their own districts.

Now it was time to govern. It was time to shut down our first ten deadly open-air drug markets as we had promised. It was the critical first step in our drive to make our city safer.

And CompStat would be our method.

Enter Jack Maple

The first time I spoke to Jack Maple was the day after my own Election Day. As luck, or Providence, would have it, he and his business partner, John Linder were in Washington, DC—just an hour down the road. They agreed to meet me for dinner that night at a little Italian place called Maria's on Connecticut Avenue.

After a long dinner that night, Jack agreed to come help Baltimore but with one confidential caveat. He was fighting cancer. The prognosis was not good. And he could only be involved for as long as his health allowed.

As fate would have it, I'd be the last mayor to benefit from Jack's compassion, expertise, and counsel in reforming a police department. And the first mayor, with Jack's help, to take CompStat enterprise-wide for the whole of city government.

Thanks to philanthropic business and foundation leaders, we were able to raise the money necessary to hire Maple and Linder and put them to work in Baltimore. As Maple assessed data systems and extracted hard counts of deployment and sworn strength and equipment, Linder went to work conducting dozens of focus groups across every imaginable slice of the Baltimore Police Department—young officers and old officers, sergeants and lieutenants, male and female officers, black and white officers. It was the first time that the clear majority had ever been asked their opinions about crime-fighting and the strengths and weaknesses of their own police department. But it would not be the last time.

All this intense listening would be used to design a polling questionnaire that every member of the department would be asked, but not compelled, to complete. The goal was to give every member of the

Police Department some ownership of the change and reform that was coming—a sense that they were recognized and heard, and even helping to shape the improvements we all wanted to see.

The plan was for the new police commissioner to send the polling questionnaire out to the department over his signature. The wording of every question—based on language and substance gleaned from the focus groups—would communicate that their new leader felt their pain and understood their hopes, desires, and frustrations. The results of the poll would be made public for all to see. And the reformation would begin.

We vowed to one another never to relent, and never to let up. This would be a long, uphill push.

Building Out a Proper CompStat Room

It is said that form follows function. And this is certainly true when it comes to the CompStat Room (capital R intentional.) It is also said that expectations become behavior. And there is an expectation created by a proper CompStat Room.

Soon hammers were swinging, and drywall was going up over at the fourth floor of the Police Department headquarters building. Under Maple's supervision, a new and permanent CompStat Room was under construction. Gone were the days of the occasional *faux* CompStat, where once a month a projector was slapped on top of a card table in the police auditorium. Faux CompStat was little more than a slide show of numbers without honest dialogue or discussion about how to solve a murder, shooting, or armed robbery problem. It was a dog-and-pony imitation of the real thing.

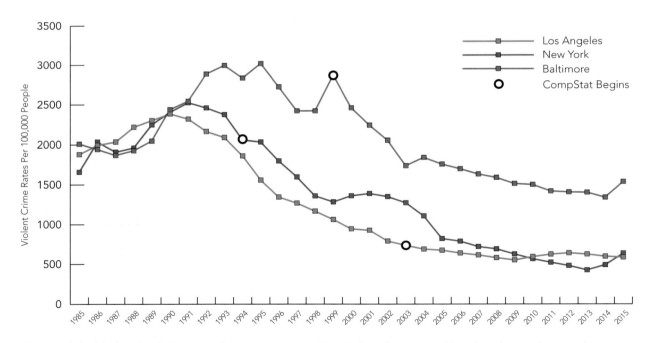

This graph highlights the decline in violent crime rates in New York, Baltimore, and Los Angeles, and notes when CompStat was introduced in each city. It also shows crime rising again in Baltimore as city and police leaders stopped using CompStat.

We spared no expense in building out a new CompStat Room in the newest section of the Police Department headquarters. And everything about the new CompStat Room said permanent—a permanent array of desks in a horseshoe formation for the citywide police command staff; permanent projectors mounted in the ceiling; a permanent glass-enclosed booth for the staff from Planning and Research to follow the conversation and project the corresponding maps, charts, and graphs; permanent big-screen monitors mounted on the walls; and a permanent podium up front where the district commander flanked by his district command staff would present—the focal point, really, of every meeting.

This would be the collaborative nerve center of Baltimore's new crime fight. And everything about that room said we were in it to win it, and there would be no turning back.

Another seed planted.

That CompStat Room also became the prototype for the new CitiStat Room that was soon under construction on the sixth floor of City Hall. The space was personally chosen by Jack Maple after touring every square foot of City Hall.

In the year that followed, several other departmental stat rooms would also be constructed in other buildings, ushering in a new way of governing across the entire City of Baltimore.

When our federal government helped us hire an additional 200 police officers, we had a decision to make. Do we divide the additional officers equally among the six City Council districts, or do we send them to small areas where the greatest numbers of our citizens were being shot, robbed, and mugged year in and year out? We went with the second option, and put our city on a path for the biggest ten-year crime reduction of any major city in the US.

Closing Open-Air Drug Markets

The leading promise of the campaign for mayor was that, together, we would close ten open-air drug markets within my first six months in office. In hindsight, such an undertaking seems quaint and small. But for a city that had seen these open-air drug markets expand unabated for twenty years, it was, at the time, a bold and crystallizing moment of accountability. None of the other candidates could bring themselves to make such a seemingly impossible commitment.

If you believe you can or you believe you can't, you are probably right. So, we chose to believe we could. But there is a big difference between saying it and doing it.

We put together the plan for closing these first ten open-air drug markets in a way that was both collaborative and data-based. We reached out broadly and brought several different groups of stakeholders into our expanding circle of decision-making. Many of the people involved in formulating the plan and determining the metes and bounds of the first areas to liberate were from inside government. Many others were neighborhood leaders and activists from outside of city government.

The one thing all of us had in common was a love for this valuable piece of the map of the United States called the City of Baltimore.

In fact, almost every one of these conversations happened around a map. And on this map, staring us all in the face, were plotted the requests for police service—the 911 calls: for open-air drug dealing, for drug dealing from within an address, for shots fired, for murder, for robbery, and for every other type of 911 call.

The map told a story. And our story was that we had let open-air drug markets operate with impunity

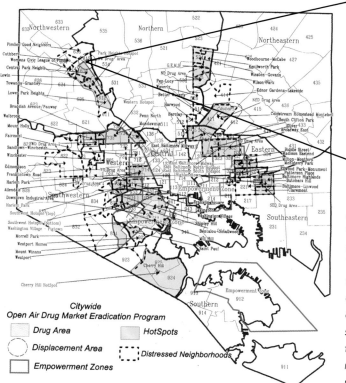

Citywide
Open Air Drug Market Eradication Program

- Drug Area
- HotSpots
- Displacement Area
- Distressed Neighborhoods
- Empowerment Zones

On June 21, 1999, I announced my candidacy for mayor promising that within six months, together, we would close down ten of our city's most notorious open-air drug markets. And we did. The difference between a dream and a goal is a deadline. The map on the left provides an overview of the areas within our city with the highest concentrations of crime. Inside these areas, the smaller blue zones mark the boundaries of the first ten open-air drug markets we set out to close. The map above is an enlargement of an open-air drug market we closed within the Northwestern District.

over whole swaths of our city. And the common denominator of the neighborhoods we had collectively written off was that they were all poor, and most of them were black.

But the map also told us that the areas we had to reclaim were not massive and boundless. In fact, when compared to much larger cities like New York and Los Angeles, they were compact and finite with easily identified borders. And even within neighborhood borders, certain hot spots became clearly visible when calls for service from citizens were mapped for everyone to see.

The designation of the boundaries would not be our most advanced use of data-driven decision-making. But it was the best we could do at the time. Neighborhood leaders from every part of the city generously came together in evening meetings to look over the maps and the data showing where various crimes were happening. Each map was ground-truthed with community input, experience, and knowledge. And some boundaries of these initial areas were also determined by hard-to-measure factors like the level of committed neighborhood leadership support.

Some in the Baltimore Police Department wanted more time to get the boundaries of these first ten areas perfect, or to get our preparation just right. But we didn't have more time. The people and I had agreed that we would close the first ten open-air drug markets within our first six months together. I was intent on keeping my part of that bargain.

We were elected to govern right now, not later. The people elected us to deliver results. We had six months to make the difference we had promised. And we would have to trust that the public would be smart enough to recognize progress even if—in some ways—it fell short of the goal.

What we needed was a beachhead. Geographically and politically.

"Was this about politics or public safety?" some would ask.

The answer was: "Yes."

It was about politics *and* it was about public safety—it was about the future of our city. Murders were now higher this year than the prior year.

We had to begin. And we had to see with our own eyes that we could succeed for a change.

A "Whole of Government" Approach

Ours was a "whole of government" approach. And it would become a whole city approach.

Eradicating these first ten open-air drug markets became every department's business. Solid Waste, Transportation, Housing, and Health. Reducing crime—and the conditions that lead to crime—was everyone's business.

In every step of the work, from the initial setting of boundaries for heightened attention and enforcement, to the initial walk-through, to follow-up inspections, neighborhood leaders were invited and encouraged to accompany city and police officials every step of the way.

Complaints about burned-out street lights of course came in from all over the city. We prioritized for the fastest service those complaints that came in from neighborhoods hardest hit by crime. We did the same with illegal dumping, abandoned cars, dead trees, and vacant houses. Instead of hoping for the complaints to be called in by demoralized and long-suffering citizens, we sent inspectors in as collaborative teams from multiple departments—during the day, and at night—to generate work orders to address these quality-of-life issues more pro-actively.

Selecting the initial ten areas took a lot of internal debate and hours and hours of meetings and honest discussions with neighborhood leaders. Every hard-hit neighborhood in the city—and there were many—wanted to be in that first group of ten. Not a single neighborhood asked for less police attention or less enforcement.

Some counseled that we should not publicize the drug areas we were working to close until after we had succeeded. Others counseled that we should have fifteen or twenty "secret" areas that we set out to close and then share the most successful ten with the press when the six-month deadline arrived. Cute.

"How will that help us motivate neighbors, churches, and businesses in the areas we are fighting to reclaim?" I asked.

Some of my staff were afraid. But they were afraid of the wrong things. They were afraid I would be hurt politically if we failed. After all, no one else had succeeded at this for decades here.

"Why will police officers risk their lives," I asked, "if their civil leaders don't have the guts to risk their own precious political capital?"

The press openly doubted the sincerity of my commitment. After twenty years of seeing so many of our neighborhoods go from bad to worse, it was hard to blame them. Some of them seemed anxious for us to make news by failing. And they knew we would fail because "all the smart people" knew it could never be done, here.

On February 1, 2000, we held a press conference to announce the first ten open-air drug markets that we would close. The location we chose for the press conference was Harford Road and The Alameda—the same corner where I had announced our campaign for mayor months earlier.

We put forward both the plan and the map. The how and the where. Larger maps of each designated

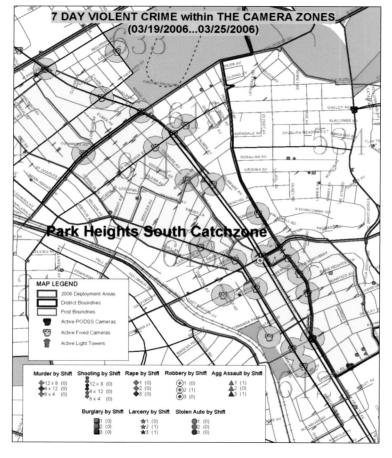

A CompStat map showing a seven day snapshot of crime in the Park Heights of Baltimore area in March 2006. Multiple overlays include the locations of cameras designed to detect crimes in the deployment focus area at that point in time.

Governing for Results: Making Progress on Drugs

The open-air drug markets were the relatively tiny square blocks of our city where drive-by shootings and retaliatory homicides happened with the greatest concentrated frequency, year in and year out. A national television series had even been aired, portraying the seeming hopelessness of ever recovering peace on these notorious Baltimore corners. So, we vowed to close them.

On Thursday, June 8, 2000, *Baltimore Sun* writer Peter Herman acknowledged that violence and drug deals were declining in targeted areas across the city:

> "Baltimore Mayor Martin O'Malley's pledge to reclaim 10 drug-infested areas within six months of taking office has been largely fulfilled, police said yesterday, with crime down and fewer people complaining about dealers and addicts.

> "Homicides and shootings also dropped on streets surrounding the designated drug markets, which police say shows they are not simply shuffling the drug trade from one block to another.

> "'The liberation of Baltimore's neighborhoods has begun,' O'Malley said yesterday while standing at North Rose Street and Ashland Avenue, ground zero for a band of frustrated residents who have confronted dealers."

Even the ultra-skeptical *Baltimore Sun* had something half-encouraging to say about this new "effort" in Baltimore's fight against crime. Though, understandably, with murders still up, they could not call it "progress." In a June 11, 2000, editorial, they wrote:

> "After the first six months, the jury is still out on the mayor's effectiveness. He has raised expectations by setting goals and making announcements. But he has yet to show major achievements. While the mayor has made good on his campaign promise to shut down 10 of the worst open-air drug markets, drug-related killings have continued unabated. At the current rate, this year will again end with more than 300 homicides—for the 11th consecutive time.

> "The mayor's vigorous style, nevertheless, has produced a sense of hope unfelt in Baltimore for more than a decade."

We were making progress. But, we had more work to do.

area were displayed on easels for the press to poke at. These were posted online, as well. The entire City Council was briefed. We itemized tactics like street closures, civil citations, enforcement of quality-of-life crimes, field interviews, publishing the names of buyers caught in undercover reverse sales, and more. And importantly, we put in big bold print, "No action will be taken at the expense of another's civil rights."

"Whereness," the author John O'Donohue wrote, "is crucial to human identity." Place matters. Together, we were affirming by our commitment and the actions to follow, that in our city, every neighborhood matters and every place matters.

We identified the first ten areas to be reclaimed from open-air drug dealing and went to work.

The questions from the press that day reflected the cynicism of the times and the culture of failure that had allowed our city to become the most violent, addicted, and abandoned in America.

Question: "How is this effort going to be different from the usual Drug-Free Zones that we've seen designated for years?"

Answer: "We are going to succeed this time... next question."

Question: "Won't you just move them from one area to another?"

Answer: "Not if we lock them up on proper charges and the State's Attorney's Office secures good convictions."

Question: "But they don't stay in jail; what happens when they get out?"

Answer: "If they return to deal drugs openly on our streets, we are going to lock them up again, and ask the courts for a stiffer sentence on the next offense. You bring the noise, you get the noise—welcome to America."

Question: "Why did you pick such a small area in this district (implicitly, an area so small that you cannot fail)?"

Answer: "Based on a number of factors including community support and the best judgment of district majors."

Question: "What happens if you succeed at closing down these open-air markets, but it just moves someplace else?"

Answer: "Then we'll hold what we have reclaimed and go to the other neighborhoods and shut down the drug dealer's operations there, too. Thank you all for coming..."

The birds in the rafters were poised to crow "failure."

Our city government and our Police Department, however, were now poised to succeed.

Small Things Done Well Make Bigger Things Possible

The next few months were some of the most important in all the years of crime reduction that would follow. They were also some of the most important months in our city's struggle to recover our lost sense of confidence—to recover our true self.

Our newly appointed police commissioner moved quickly. He led the department forward with CompStat and countless visits to roll-calls. He led from the front and drove CompStat from the center. He asked every day whether the Police Department was following through with tactics and strategies to reduce crime, and asked whether those tactics and strategies were working. He backed up police officers when they were right, retrained them when they made honest mistakes, and showed a decisiveness to charge and prosecute them when they betrayed the public trust.

As mayor, I maintained a steady day-and-night schedule that had me returning time and again to the hardest-hit neighborhoods of our city, interacting in meeting after meeting with neighborhood leaders

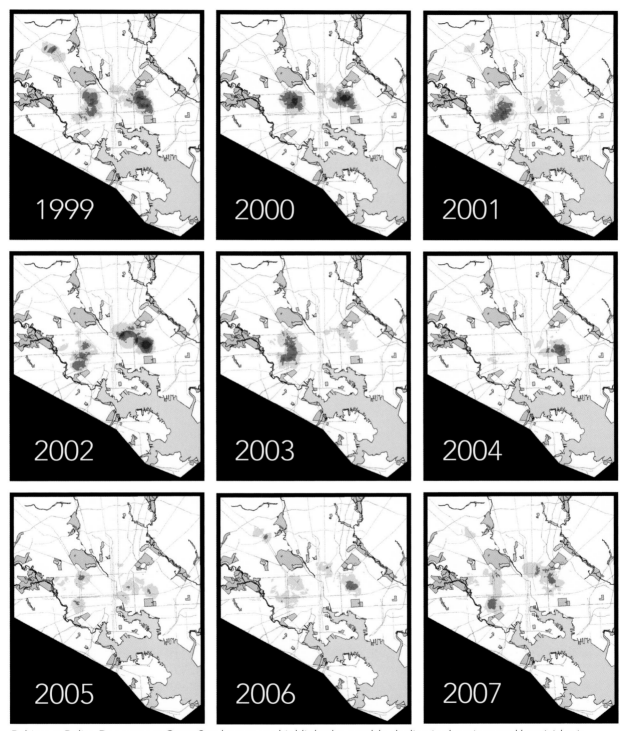

Baltimore Police Department CompStat heat maps highlight the notable decline in shootings and homicides in Baltimore from 1999 to 2007.

who were on the front lines of our fight to make Baltimore a safer city. In large public meetings, I routinely gave a pitch for the newly staffed Civilian Review Board. I explained its oversight purpose under city ordinance to act as an independent investigative body for complaints of police discourtesy or excessive force. I would read out the phone number for Civilian Review, slowly. Twice. Whether it was "Mayor's Nights Out" in a town hall fashion, or "Mayor's Nights In" with community leaders inside City Hall, or numerous other events at churches or neighborhood cookouts, everybody knew if they had a problem with policing (or anything else), they could talk to the mayor. And he would be back. Neighbors knew it. And so did the police.

For all the new technology for analyzing and mapping crime and crime trends, much of the work in improving the effectiveness of our Police Department came down to a focus on doing a few leading actions well.

Small things done well make bigger things possible.

Fortunately, there were many highly motivated, honest, and courageous officers still within our department who were eager to be well-led. They knew that we had been losing this fight for years. And they were ready for a new day. Most of the Baltimore Police Department wanted to win, and many still believed we could. The leaders within the department would need to be identified and lifted up. Many would self-select. The proof would be in their performance. Performance that everyone in the department—thanks to CompStat—would be able to see for themselves and follow.

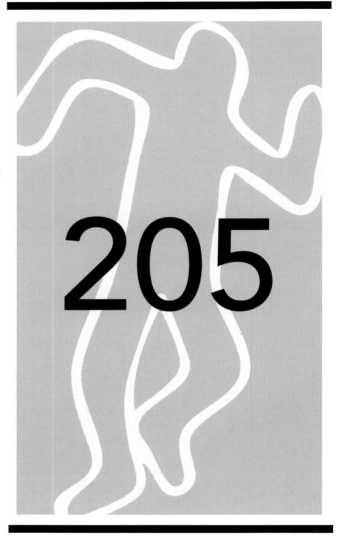

"Chalkman," the grim performance metric on homicides that the Baltimore Sun started running on its op-ed page every week beginning in mid-2000. The newspaper stopped running Chalkman when we reduced homicides to record lows.

When the six-month deadline for closing the first ten open-air drug markets arrived on June 7, the public had the opportunity to judge for themselves and the reporters didn't want to write the story. In fact, they looked for ways not to write it. Maybe they just didn't believe they could ever write it, but the numbers told a compelling story of progress.

The news editors sent their young reporters (we had both back then) out to canvass the neighborhoods where the drug dealing had been rolled back. They hoped the neighbors would unite in denouncing our achievement, but they didn't—most neighbors saw the difference and appreciated the effort. They saw progress where there had been none for years.

This was not a "mission accomplished" moment. It was, however, the successful start of a long-needed mission of recovery. This was Representative Democracy 101—performance management in action. And not just the new mayor measuring the performance of city staff and departments, but residents of the city measuring the performance of the new mayor in their own neighborhoods.

Chalkman

Around this time, the *Baltimore Sun* introduced a new graphic on its op-ed page. We called him "Chalkman." Every Friday, there now appeared a chalk-outlined figure of a murdered body sprawled on the pavement. On top of the body's chalk outline were the total number of homicides for the year to date. Underneath was the comparable number of homicides from the year to date before.

"God bless their cynicism," Jack Maple chortled. "At least, now, they are playing along."

Some of my senior staff resented Chalkman. The *Sun* had not printed Chalkman under any previous mayor. I told them we should welcome it. They were going to hold me accountable to a promise they felt I could not keep, but we had volunteered to be accountable to more than their ink. And we did not intend to fail.

By the end of calendar year 2000, we had driven homicides below three hundred for the first time in twenty years. And we had driven every other crime category down, as well. The open-air drug markets that were closed, remained closed. We started opening inpatient drug treatment centers for the first time in thirty years. We persuaded the state to double, and then quadruple, drug treatment funding. We began to intervene more effectively in the lives of young people at risk, and vulnerable families who needed help. And although we didn't know it at the time, that first difficult year set our city on a course for the biggest ten-year crime reduction of any major city in America.

We did not make our city perfect. We did not make her immune to setbacks or future crises. But we did save a lot of lives.

And a city that had been shrinking for thirty years began to grow and breathe again.

Fairness, Accountability, and Collaboration

There were a few times during the campaign of 1999 when national reporters would write a scant article on the mayoral contest in Baltimore. They would usually pigeon-hole my candidacy as the "white candidate promising to crack down on crime." In their box of "the ways things have always been," they would conclude that being white and promising to crack-down on crime meant I didn't have a chance in a black city. But what they missed was that I was talking about something larger and deeper than "cracking down on crime."

I was talking about *justice* and *injustice*. And I was asking the people of our city if we were willing to do something about it.

I was talking very directly about the injustice of watching 300 young black men shot to death on our streets year in and year out as if there were nothing we could do about it.

I was talking about the injustice of allowing open-air drug markets to operate in poor neighborhoods when we would never allow that sort of lawlessness and violence in wealthier neighborhoods, black or white.

I was talking about the injustice of having one standard for law enforcement in wealthier neighborhoods, whether black or white, and another standard in poorer neighborhoods—most of which were black.

I not only promised—if elected—to improve policing; I promised to improve how we police the police, train the police, recruit the police, and when necessary, discipline, fire, and/or prosecute the police.

It was the understanding that the two goals must be pursued together—that effective policing and fair policing reinforce one another and make each other possible—that created the public consensus and the public mandate to try.

Then, as now, every honest discussion about improving policing in Baltimore also involved a discussion about how to guarantee that policing is done in a fair way—with professionalism, courtesy, and respect for citizens; without racial profiling, discourtesy, and the use of excessive force. This was the conversation throughout the campaign, and this was the hard work in every day of the governing that followed.

As hard as we worked, the day never arrived when we thought, "OK, we've accomplished the mission—we are policing as effectively and as professionally as we ever can. We can now move on to other things..." It took constant work and tending (even as we tended to other things like schools, health, jobs, and housing at the same time). It took both intention and action. Thoughtful planning and relentless follow-up. Humility and a passionate drive. And always the very human dynamic of identifying and lifting up the best leaders within the department.

This was not a matter of checking a box and moving on. It was more akin to the ongoing work of planting and tending a garden. If you sow the seeds of justice, you will reap the fruits of peace.

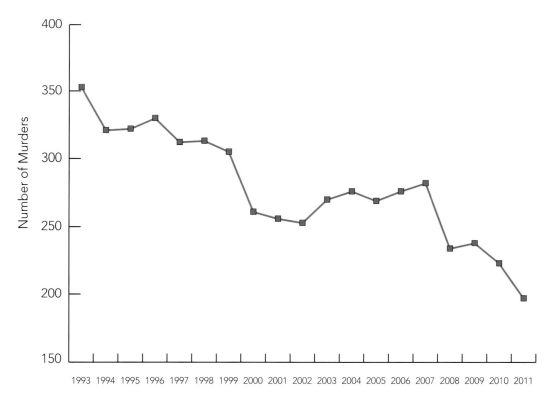

Decline in the number of murders in Baltimore, 1993 to 2011.

Leading Actions to Reduce Crime

At the conclusion of Jack Maple's book, *The Crime Fighter: How You Can Make Your Community Crime-Free*, he lays out a "canon for crime-free communities"—a series of actions, strategies, and policy changes that, if taken, would make all neighborhoods in America virtually crime-free.

If the list were read aloud, it would make law enforcement heads nod as if to say, "We are already doing that." But the reality is that every police department is different in terms of the things that they do well, the things that they do poorly, and the things they think they are doing well but really are not doing well if at all.

How sad it is that drug dealers implement new ways and new technologies so quickly, but police departments...not so much?

When we first started, measured against the points in Maple's canon, there was very little the Baltimore Police Department was doing well. In fact, there were many things we weren't doing—or even trying to do—at all.

"We are in last place," Jack would frequently remind us. But with his guidance, we started doing some basic things right. We began to focus on the core leading actions that every police department should take to reduce violent crime.

We started holding subordinate commanders accountable for the reduction of crime in their areas and gave them operational authority to synchronize the work of patrol, detective, and narcotics squads in their districts.

We started paying our police officers and their supervisors more so that we could better retain experienced supervisors, and better recruit a more diverse new generation of police officers. (We negotiated a three-year contract with 7 percent, 8 percent, and 9 percent salary increases, as well as step increases based on seniority to retain experienced supervisors and commanders.)

Instead of acting as if the initial arrest were the end of crime-solving, we started interviewing and debriefing all arrestees about other open crimes in their area. Improving the rate of cases solved became a daily pursuit, not a fact to be glossed over in year-end wrap-ups.

We created, or re-created, a cold-case squad to revisit stalled homicide and shooting investigations.

We created a metropolitan fugitive task force—with help from Baltimore County—to go after the most violent fugitives in the backlog of 50,000 open warrants—warrants that used to be handled by just four detectives.

We created "flex squads" that could be deployed to suppress crime proactively based on where we saw it emerging, rather than reactively responding to 911 calls after the crimes had been committed.

We created an expectation—even with the occasional setbacks beyond our control—that every two weeks crime should be driven down, and every two weeks more cases should be solved. Progress, like interest, compounds.

We began changing our Police Department into a much more transparent organization by regularly reporting crime statistics and posting the crime maps online for every citizen to see and analyze for themselves for the first time. And importantly, we did this in ways that made it easy for citizens to see, not harder; we posted maps of where, when, and how real people were hit by crime rather than simply posting impenetrable tables of numbers.

We started giving crime victims same-day service—or as close as possible to same-day service—to sit with a detective for an interview or to view photos of possible suspects.

We elevated the handling and preparation of gun cases—particularly for repeat offenders. And we began to very vocally and repeatedly ask the State's Attorney and the federal US Attorney's Office for greater effort in the prosecution of gun violence.

Most importantly, we started managing deployment, crime suppression and apprehension, and crime-solving the CompStat way. We started using Jack Maple's four management tenets: timely and accurate information shared by all, rapid deployment of resources, effective tactics and strategies, and relentless follow-up and assessment.

Every week, CompStat gathered district commanders and their staffs as each district came through police headquarters to present on a rotating basis. The questions were asked: What are we doing that is working, and how can we do more of it to reduce the numbers of our citizens who were being robbed, mugged, shot, raped, or otherwise victimized? Ideas and tactics were shared during the meeting and just as importantly after the meeting—in the hallway among the detectives and officers who were doing the work.

Communication, coordination, and cooperation—usually unnatural acts among non-consenting adults—were made to happen routinely. The achievements of leaders within the organization—whether patrol, detective, or narcotics—were recognized not just by their commanders but by their colleagues.

The police commissioner made it clear by his time and presence that this was not a passing fancy. This was the new way of getting things done and saving lives in the Police Department. The compelling scoreboard that everyone could see was crime reduction. A cadence of accountability was created by the frequent, routine, and predictable schedule of regular CompStat meetings.

As mayor, I made a point of popping into a meeting every other week or so—usually unannounced—to sit next to the police commissioner; to listen, to see, and to observe. On the rare occasion that the conversation led me to want to ask a question, I would whisper it to the police commissioner rather than ask it myself. I was there to reinforce him, not to undermine him. And I was always careful on each visit to make sure all my body language and demeanor conveyed my trust and confidence in the commissioner.

Leading Actions for Policing the Police

There are certain aspects of policing that to the casual observer might not seem essential to the crime fight. But these actions go directly toward building the trust necessary between police officers and the communities they serve. Without trust between citizens and police, no amount of policing can be successful in the long run. And all these actions were within the power of the city and its Police Department to carry out.

All of us as Americans have inherited a legacy that we must acknowledge and address. It is the legacy of slavery—a legacy that has been intertwined with racial injustice and law enforcement from the first days of our country's colonization and founding.

We can pretend that there is nothing we can do about violent crime, police brutality, and racial injustice in its many forms; or we can learn the lessons of history and take actions to improve the future.

Progress is a choice.

Among the areas to which we gave a heightened level of political will and management attention were Internal Affairs, civilian review, reverse integrity stings, excessive force and discourtesy complaints, and early warning systems.

- Internal Affairs: An effective Internal Affairs Division (IAD) must have sufficient sworn strength to follow-up immediately on complaints of police misconduct, discourtesy, and excessive force. Interminable investigative backlogs and trial board backlogs greatly undermine police integrity efforts. We began to increase the staffing and the caliber of leadership within the Internal Affairs Division.

There are many experienced police leaders who believe that 5 percent of any police department's staffing should be dedicated to Internal Affairs to properly protect the integrity of the force.

- Civilian review: Civilian review serves a vitally important function as an independent oversight entity to which citizens can turn with trust and confidentiality. Our Civilian Review Board first came into being by city ordinance (which I had co-sponsored as a council member) just shortly before the 1999 campaign began. But appointments to the Civilian Review Board must always be filled with good, reputable, and committed community leaders, and to properly function it must meet regularly. For the first time, we gave the Civilian Review Board the budget authority to hire its own independent detectives. An effective Civilian Review Board must have the freedom and capacity to thoroughly investigate the cases it deems worthy of further investigation in a timely manner.

- Reverse integrity stings: A reverse integrity sting is an undercover operation to test whether police officers do the right thing or the wrong thing when they think no one is watching—like planting drugs on a citizen to clear a complaint. During the campaign and in the governing that followed, I promised that our Internal Affairs Division would start conducting one hundred reverse integrity stings a year to proactively safeguard the integrity of our police force. When an officer failed one of these tests, the political embarrassment was always outweighed by the increased confidence that citizens felt in knowing their government was serious about policing its police force.

- Excessive force and discourtesy complaints: We began tracking and sharing the volume of these complaints—citywide and by district—and we started proactively sharing the totals, subtotals, and trendlines regularly with the City Council and with the public. Every day we worked to improve police professionalism on the streets, as well as training, and supervision. We worked to drive excessive force

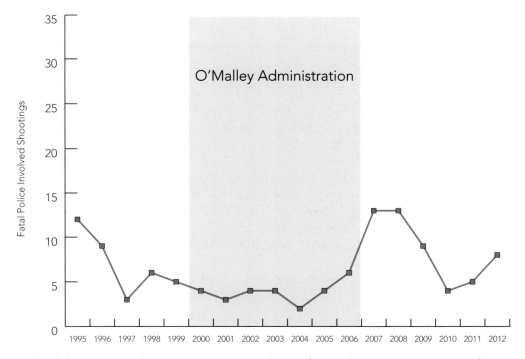

Fatal police involved shootings in Baltimore, 1995 to 2012. Police professionalism, courtesy, excessive force, and use of deadly force can all be measured and mapped. With the right combination of actions, Baltimore can take steps to increase police professionalism and courtesy and reduce the use of excessive and deadly force.

and discourtesy incidents down every year. We succeeded in driving police-involved shootings to the seven lowest years on modern record. As professionalism and supervision improved, so did trust between citizens and their police force.

- Early warning systems: For the first time, we implemented IADstat, a specialized version of CompStat focused exclusively on issues of police integrity and professionalism. We began, also for the first time, to identify the police officers who were racking up discourtesy or excessive force complaints in great numbers over short periods of time. This early warning and follow-up system worked. Officers drawing high numbers of complaints would be pulled from the streets, so we could find out why. Sometimes, high-performing officers were drawing complaints from drug gangs who were trying to sideline them. Other times, the officer required retraining or other effective interventions due to personal issues. At times, tighter supervision or greater discipline was warranted.

The Future of Policing in America

Honest, fair, and effective policing is a public good of the highest order. It is also the indispensable center of the drive for a more effective criminal justice system—a system in America that cries out for reform.

There are many functions of our criminal justice system in America that mayors cannot control, and there are many levels of government and branches of government that mayors cannot change or affect. But policing is one they can.

Policing is the essential public function that links public trust and public safety. It is the human element that connects trust and safety over the course of tens of thousands of individual interactions across our country every single day. While some of those interactions are casual and pleasant, many are tense and confrontational. Many are dangerous, some are deadly.

A prayer vigil in front of the burned-out home of the Dawson family on October 17, 2002, immediately following one of the worst violent massacres in our city's modern history. In retaliation for standing up against drug dealers in their own neighborhood, Mr. and Mrs. Dawson and their five children died in a horrific fire after a drug dealer kicked in their door in the middle of the night and threw a Molotov cocktail into their living room.

It is important to work on the root causes of crime—poverty, hopelessness, gross inequality, broken homes, lack of jobs, lack of education, lack of housing, lack of so many opportunities.

But without public safety, civil society ceases to exist.

When I was elected governor in 2006, we finally extended the reach of reform beyond policing to some critical public safety agencies of state government—Parole, Probation, Corrections, Juvenile Justice, Child Protective Services, and Foster Care.

By using data, the map, and proven methods of performance management, we were able to create far more effective partnerships with local law enforcement than ever before. With a constant focus on violent crime reduction, we also implemented a host of criminal justice reforms. We repealed the death penalty. We decriminalized possession of marijuana. We restored voting rights to 75,000 people who had served their sentences and returned to society. With a focus on more effective ways to manage re-entry, we also reduced recidivism by 20 percent and reduced Maryland's incarceration rate to twenty-year lows. By 2014—the end of my two terms as governor—we had reduced violent crime in Maryland to thirty-five-year lows.

During my service as mayor of Baltimore, courageous neighbors and their police officers drove violent

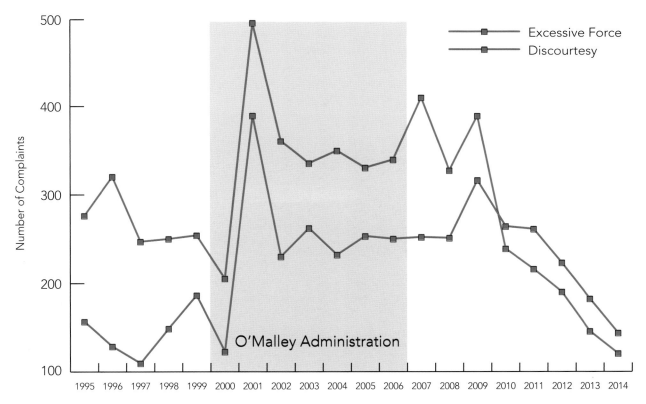

Graphs are important but graphs can also be open to interpretations not always supported by reality. In 2000, we made a concerted effort to enhance efforts at policing the police, and actively encouraged neighbors to report unprofessional, discourteous, or abusive behavior. In the three years prior to the riot that rocked Baltimore after Freddie Gray's custodial death, those efforts had been de-emphasized and in some instances—such as civilian review and reverse integrity stings—they had been discontinued.

crime down by 36 percent. It was not the 50 percent reduction we had strived to reach, but it was the biggest reduction in crime Baltimore had achieved in decades. And the people who experienced the biggest positive difference were the people of our city who lived in our poorest neighborhoods.

For a while, that positive, life-saving trend in Baltimore continued. My immediate successor kept many of the policing reforms, like CompStat and civilian review. By 2009, we had driven annual homicides down to just above two hundred. A majority white police force had been affirmatively recruited and trained to become a majority African American police force. And from 2000 to 2009, Baltimore led all major cities in America in the rate of reduction of crime—an achievement that went largely unreported locally.

But leadership is the great variable. And by 2011, new leaders were in place in Baltimore City Hall and in the Baltimore Police Department. And these leaders somehow lost focus on the constant work of improving and adapting tactics and strategies. CompStat meetings became less and less frequent until they stopped happening at all. Reforms to improve the professionalism, training, and integrity of the department also fell by the wayside. Civilian review ceased to function, and reverse integrity stings ceased to happen. Citizens in our poorest neighborhoods didn't bother to make complaints anymore about excessive force or discourtesy because they knew nothing would happen if they did.

In the name of budget savings, the patrol strength of the Baltimore Police Department was reduced by hundreds of officers over successive budgets. Enforcement effort diminished—especially in our poorest neighborhoods. And, not surprisingly, by 2014, arrests had plummeted to a forty-year low. It might have been even lower than a forty-year low, but such records had only been kept for the last forty years.

Very few of these incremental policy reversals ever made the local paper. The larger story never did. But the rock was clearly starting to roll back down the hill by 2013.

Row houses on Saint Paul Street with downtown Baltimore in the background.

Then the spring of 2015 happened. Outraged by the cell phone video images of a seemingly rapid succession of black citizens being shot by police, protests and riots rocked many American cities. One of those cities was Baltimore.

On April 12, 2015, Freddie Gray Jr., a twenty-five-year-old African American, was arrested by the Baltimore Police Department for possessing what the police alleged was an illegal switchblade under city law. While being transported in a police van, Gray fell into a coma from which he never recovered. He died on April 19, 2015. The coroner ascribed his death to injuries to his spinal cord. The six officers involved in his arrest were suspended with pay. A cell phone video of his arrest was shown every night on television as protests grew. At the same time, other events involving the loss of black lives at police hands unfolded across the country. And then two weeks later, on April 25, Freddie Gray's West Baltimore neighborhood erupted in a single night of angry fires and protest.

Unlike Los Angeles and New York, the drive for a safer Baltimore has never really recovered from Freddie Gray's tragic death and the lack of effective leadership that preceded it. Sustained improvements in public safety require leadership that is willing to change, adapt, and take relentless actions that improve public trust in policing. This requires openness, transparency, relentlessness, and a demonstrated commitment to policing the police.

Criminal justice in America is a system. It has many independent and interdependent parts. But it need not be an unintelligent system, a wasteful system, a brutal system, or an uncaring system.

Many different political authorities, and many different branches of government—executive, legislative, judicial—play a role in improving the effectiveness and the fairness of this human system.

And while police reforms alone are not enough to sustain long-term improvements to public safety, no amount of other reforms can make up for the absence of fair and effective policing.

We know what works to improve public safety and criminal justice in America.

With effective leadership, the Information Age provides the tools.

It is time to learn from what works and do it.

Learn & Explore

New York's Gospel of Policing by Data Spreads Across the United States
Read an article about the spread of NYPD's CompStat to Baltimore and other cities.

Exploration of Criminology Data
Police departments can use dashboards to let citizens explore criminology data by police division, district, crime category, type, or a specific date.

Neighborhood Officers
The City of San Luis Obispo, California, created this story map to help citizens connect with the officers assigned to their districts.

For links to these and other examples, exercises, and resources, visit SmarterGovernment.com and click chapter 5.

CitiStat and the Enterprise of Governing

CompStat changed policing in the United States. The data, the map, and the method allowed police commanders to deploy more rapidly than they ever had before. The repeatable cadence of biweekly meetings focused the collective experience and knowledge of the group on solving problems, running plays, and getting inside the turning radius of the problem—the problem of crime.

Could these same principles and practices be used to run the entire enterprise of city government?

We decided to find out.

Taking CompStat to Scale

Baltimore was not the first city to borrow CompStat to improve public safety. But we were the first city in America to apply the tenets of CompStat to the entire enterprise of city government.

We called it *CitiStat*. And it was combined with another new technology—the use of a single phone number, 311, for all citizen service requests. CitiStat earned the Innovations in American Government Award in 2004 from the John F. Kennedy School of Government at Harvard University. *Governing Magazine* in 2001 wrote that Baltimore was managing performance "on a scale never seen before in local government." Since that time, cities in the United States and around the world have taken these new tools and practices to ever-high levels.

Smarter governing has given rise to smart cities. Today, almost every big city in the United States uses the data, the map, and some method of performance management somewhere in their organization. Almost every big city uses a single phone number for all citizen requests for city services. Even so, some cities are better run than others, and leadership remains the great variable.

In Baltimore in 1999, the technology and the method were all very new: measuring deliverable outcomes every day instead of budget inputs every year, driving performance, mapping problems, as well as delivery of services, creating a culture of collaboration, lifting up the leaders.

For decades, visitors would come to Baltimore to see how our city reclaimed a decrepit waterfront and turned it into a tourist attraction. As our administration picked up steam, mayors and mayoral candidates from other cities started coming to Baltimore. They were not coming just to see the reclaimed Inner Harbor. They were coming to see and understand CitiStat. Our distinguished American visitors included public administrators from state, local, and federal governments. And frequently, we had visitors from other nations—shepherded sometimes by our US State Department or Vice President Al Gore's National Partnership for Reinventing Government.

Who knew picking up the trash could be so fascinating?

The Birth of CitiStat

CitiStat was born in a Chevy Blazer rolling through the neighborhoods of East and West Baltimore.

I don't remember the exact date. But I do remember the Chevy Blazer. It had been my rolling home and office for the prior few months. It belonged to Dennis Dwyer, a friend from my boyhood days. Dennis had very kindly taken time out of his life to drive my staff and me through every day of a grueling mayoral campaign in 1999.

By early November, the primary and general elections had been won. The campaign was over. The wild world of governing was coming upon us fast. There was no time to waste. And we were determined to hit the ground running.

On a gray day in early November, Dennis and I picked up Jack Maple at Baltimore's Penn Station. He was in town for what would be several visits in advance of my official swearing-in as mayor. Jack wanted to see and understand the city—especially the toughest parts of the city.

We immediately proceeded to drive around the neighborhoods of the Eastern District of the Baltimore Police Department. And as our inspectional tour began, Jack proposed a game; he called it "Spot the Cop." Like most of Jack's games, the game had a point. And the point was a question.

The idea was to see how long it would take us to spot our first patrol car in a hard-hit section of town where one might rationally expect to see more police rather than fewer. First spot wins. The other guy had to buy you a drink at the end of the evening. In those days, we could usually go fifteen to twenty minutes without seeing a single police vehicle. Many of the corners were thick with the deadly activity of open-air drug markets.

"We're getting lots of offers to sell us drugs, but where the heck are the police?" Jack asked.

It was a question people in East and West Baltimore had been asking for some time.

These guided inspectional tours were how I helped Jack understand the city and what life was like for people living in our hardest-hit neighborhoods. But these tours were also how Jack helped me understand how to tackle the big managerial responsibility that I had just been given. The learning had to happen fast.

There came a time on this first inspectional tour when I said, "You know, Jack—I bet you could apply CompStat to some other functions of city government besides just the police."

"Not just some, Mr. Mayor—you need to apply it to every function of city government." Jack had obviously been thinking about this for a long time. "I urged (New York City Mayor Rudy) Giuliani to do it citywide. He did a little bit with trash and a little bit with jails, but he never went all the way."

"Spot the Cop" soon gave way to another inspectional pastime—"Stump Mr. Maple." In this game, I would toss up a city government challenge and defy Jack to apply CompStat principles to addressing and solving it. The conversations—on a variety of challenges—went something like this:

"At-risk kids and kids on juvenile probation, Jack," I would posit. "They aren't potholes. They have free will, complex home environments, learning disabilities, lead poisoning. You can't just apply CompStat to saving at-risk kids."

One of many "neighborhood walks"—with neighborhood leaders, clergy, council members, and representatives of every city agency and department. When everybody knows, then everybody knows. There is no substitute for being present and seeing the truth of intractable problems on the ground.

Jack would reply, "Yes, actually you can..." And then he would launch into his mantra.

"You start by asking the right questions. Who? What? When? Where? Why? Like this—who are the kids who are 'at-risk'? What's your definition of 'at-risk'? What are the contacts with social services or law enforcement that the numbers and tragic outcomes say makes a kid at-risk? What does the data tell you about the way you've always determined 'at-risk'? What can we do to help a kid as soon as we know he's at-risk?"

He was just getting going.

"When are the kids who get shot in your city—week in and week out—actually getting shot? Is it from 9 a.m. to 5 p.m. when the juvenile probation agents are working, or is it more like 11 p.m. to 3 a.m.? Where are the neighborhoods in your city where the largest numbers of kids get shot with the greatest frequency? Is it all over the place, or just in pockets? Where are the recreation centers located compared to where the kids are harmed? What hours are the rec centers open? And why? What hours should they be open? And why aren't they?"

It became apparent to me that Jack had come to Baltimore for more than the crime fight. Even with his terminal diagnosis of cancer, he had one more project to complete. And he was passionate about it. He was here to help pioneer a new and better way of governing a city.

Toward the end of his book *The Crime Fighter*, he had already laid out his dream: "So I have another

There is no amount of new police officers a city can hire to make up for a populace than has given up trust in its own police or hope for a better future. "Believe" was our campaign to awaken Baltimore's true self and call upon the spirit of our people to confront the violence of open-air drug markets, drug addiction, and drug murders.

Governing for the Individual: "Show Me My House"

During my years as mayor, we would often invite neighborhood leaders to City Hall. I would have the opportunity to show my citizens and neighbors—my bosses—their new performance-measurement tool of the CitiStat Room with its charts, graphs, timely accurate information, aerial photography, and maps. Without exception, and regardless of whether the group was from a black or white neighborhood, a rich or poor neighborhood, my presentation was always interrupted within ten minutes by the hand in the back and that question—"Can you show me my house?"

Why is that? Is it to know that I matter to my government? Is it to know that I matter and have value to my neighbors? Is it to know that my government works and therefore matters to me? Is it to understand what is around me? Or maybe is it because of a deep, innate human instinct to better understand my relationship to the forces and people around me and their relationship to me?

Little more than a decade ago, seemingly the whole world snapped awake to the power of imagery of the earth from above. We began by exploring a continuous, multiscale image map of the world provided online by Google, Microsoft, and other companies. A combination of satellite and aerial photography, these pictures of Earth helped us experience the power of imagery, and people everywhere began to understand some of what GIS practitioners already knew. We immediately zoomed in on our own neighborhoods and saw locational contexts for where we reside in the world. This emerging capability allowed us to see our local communities and neighborhoods through a marvelous new microscope. Eventually, naturally, we focused beyond that first local exploration to see anywhere in the world. What resulted was a whole new way to experience and think about our planet.

These simple pictures captured people's imagination, providing whole new perspectives, and inspired new possibilities. Today, virtually anyone with internet access can zero in on their own neighborhood to see their day-to-day world in entirely new ways. In addition, people everywhere truly appreciate the power of combining all kinds of map layers with imagery for a richer, more significant understanding.

Initially, we zoomed in on our homes and explored our neighborhoods through this new lens. This experience transformed how people everywhere began to more fully understand their place in the world. We immediately visited other places that we knew about. Today, we continue by traveling to faraway places we want to visit. Aerial photos provide a new context from the sky and have forever changed our human perspective.

Orwellian vision dancing in my head. In that not-so-distant future, a crime fighter could take a walk anywhere in America—past schools, hospitals, parks, firehouses, across roads and over bridges—and know that each of the agencies providing those vital services has identified its primary objective and has established a way to measure progress toward that goal, a way to map that information, and a regular forum at which the agency's leaders question their subordinates in excruciating detail about their intelligence-gathering systems, their deployment, their tactics, and their follow-up."

That was his vision: elegant and practical. He had come to Baltimore to bring this dream into reality.

And that is how CitiStat came to be.

Go Big, Ramp Up Fast

We did not embrace CitiStat to win awards. We did it to survive. Baltimore had lost more population over the prior thirty years than any other big city in America. As people and businesses left, problems piled up and revenues declined. We had to find a way, right away, to squeeze savings from everywhere. There was a cost to turning around violence and addiction, and to rescuing neighborhoods and people.

With Jack Maple's guidance, we decided to go big and to ramp up fast—to apply this new performance management regimen to every department in Baltimore City government. We were still in the golden time—that period at the start of any new administration for implementing big, new management changes. The golden time lasts only for a few precious months. And we were determined not to miss our window of opportunity.

We had only one shot.

CitiStat would become the way our new administration would operate. It would be the foundation and the method of every positive change we would make in Baltimore.

I was sworn in on December 7, 1999. The CompStat Room was completed at police headquarters by mid-January. By February, we had found the most suitable location inside City Hall to build out the new CitiStat Room—the building curator's high-ceilinged space up on the top floor. General Services quickly put up wall board for the installation of a projector booth. We bolted in permanent furniture and fixtures. We mounted big, permanent projector screens on the walls. We installed computer monitors on the floor in front of the presenter and panelists. And we put blow-up versions of the service emblems, or insignia, of every major department—Fire, Police, Housing, Health, Public Works—up on each of the side walls. Everything about the room said "permanent."

With Jack's expert talent scouting, Matt Gallagher—the young executive director of our transition process—agreed to be the first director of CitiStat. The business management reports on each department—which large volunteer committees had produced over the prior two months—were guided to completion by Matt. And those reports would serve us well as we fired out of the starting gate.

We requisitioned a handful of seven staff from a variety of departments. We found that Public Works already had an Esri license agreement for GIS, which some members of the department used every day. That Public Works map would now serve the needs of every department as our new CitiStat map. It would now be the common platform for CitiStat.

Within ninety days, we held our first meeting. We started with Solid Waste (trash)—tangible, measurable, visible for all to see. Then, every two weeks, we added another department to the "one meeting—every two weeks" rotation. After Solid Waste, we added Water and Wastewater, then Transportation, then General Services. When those four bureaus of Public Works were rotating through in a regular repeatable routine and cadence, we added the other departments: Housing, Health, Recreation and Parks, and Fire.

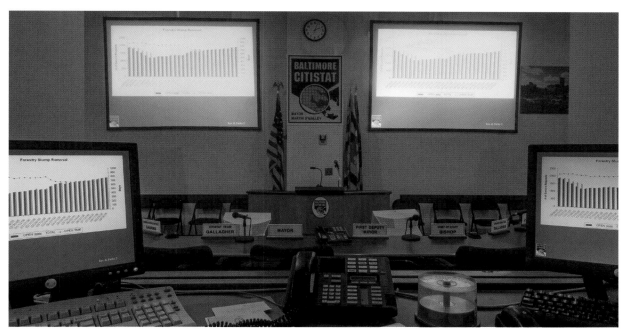

Above: The view of the CitiStat Room from the vantage point of the young staffers in the control booth who were responsible for making sure the slides, graphs, and charts flashed up on the big screens to accompany the conversations in pursuit of the truth. The stage is set for one focused hour about how we can do better in the next two weeks than we did in the last two weeks. Below: A live shot of a CitiStat meeting in progress. The mayor's citywide command staff—deputy mayor, finance director, city solicitor, director of IT, labor commissioner, director of human resources—are all around the permanently installed panel. Facing the mayor and command staff, the presenting bureau chief or department head with her command staff assembled at her back.

Implementation Timeline
for Baltimore's CitiStat

1999 to 2005

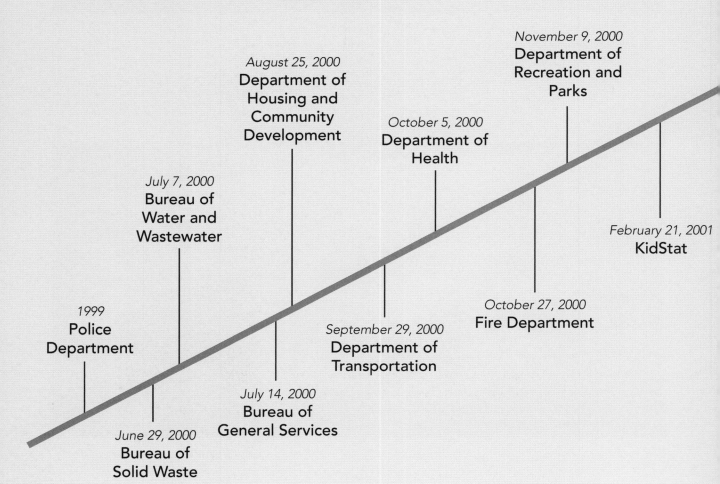

November 9, 2000
Department of Recreation and Parks

August 25, 2000
Department of Housing and Community Development

October 5, 2000
Department of Health

July 7, 2000
Bureau of Water and Wastewater

February 21, 2001
KidStat

1999
Police Department

October 27, 2000
Fire Department

September 29, 2000
Department of Transportation

July 14, 2000
Bureau of General Services

June 29, 2000
Bureau of Solid Waste

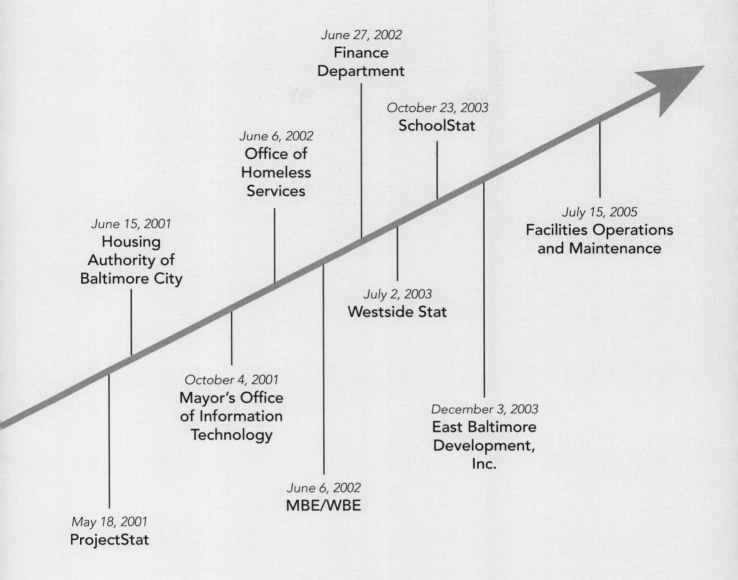

June 27, 2002
Finance Department

October 23, 2003
SchoolStat

June 6, 2002
Office of Homeless Services

June 15, 2001
Housing Authority of Baltimore City

July 15, 2005
Facilities Operations and Maintenance

July 2, 2003
Westside Stat

October 4, 2001
Mayor's Office of Information Technology

December 3, 2003
East Baltimore Development, Inc.

June 6, 2002
MBE/WBE

May 18, 2001
ProjectStat

With CitiStat, we began one department at a time until we had a rotation of ten meetings every two weeks. We started and we never stopped. As time progressed, we layered more multi-agency initiative stat meetings—again in a recurring, predictable pattern.

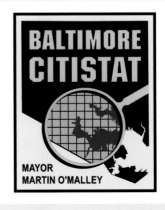

BALTIMORE CITISTAT

MAYOR MARTIN O'MALLEY

The Police Department command staff would come to CitiStat once every two weeks, as well—even as they conducted CompStat meetings in their own headquarters with a regular two-week rotation of the nine patrol districts plus Internal Affairs. Ten meetings every two weeks. Focus, collaborate, and repeat.

As time progressed, we started layering in interagency efforts like KidStat, to drive our faith-based youth violence prevention initiative; HousingStat, a collaboration between departments to deal with vacant housing; and LeadStat, which drove our initiative to eradicate childhood lead poisoning.

The Old and New Tenets of City Government

There are certain tenets—holdings of belief—that provide the foundation and the culture of any city government. And for all the cultural differences on this beautifully diverse planet, I have discovered that the tendency of governments to backslide into a culture of excuses is a universal phenomenon. It is characterized by universal behaviors. And it is given voice by similar excuses masquerading as tenets.

We set out to displace the old tenets of city government: those time-honored, well-worn excuses; the rote deflection of rational inquiry; the institutionalization of institutionalization for the sake of the institution. What follows is a short list of the old tenets of city government. It is not an exhaustive list, but just some of the more popular excuses designed to exhaust leadership and confound progress.

1. If the mayor really wants to know, we can find out, but we'll have to pull all our people off their other jobs, and it will take weeks.
2. That wouldn't work here.
3. This is the way we have always done it.
4. We tried that, and it didn't work.
5. I can't do my job because I am too busy collecting statistics.
6. We don't have the money to try that because our budget was cut.
7. I hope everybody forgets about this question before next year's council budget hearing.

We began to displace these worn-out excuses with the new tenets of city government. The tenets of CompStat rebranded—for a larger purpose—as the tenets of CitiStat:

1. Timely, accurate information shared by all,
2. Rapid deployment of resources,
3. Effective tactics and strategies, and
4. Relentless follow-up and assessment.

We started. We didn't stop. And while we had a lot of success, we never totally succeeded.

The Things That Get Measured Are the Things That Get Done

"Timely, accurate information shared by all" is always a work in progress. We didn't allow imperfect to be the enemy of the start. Sometimes the data had never been recorded before. In most cases, though, it had—albeit, often it was on paper, or on yellow legal pads, or in dog-eared spiral notebooks. But rarely did it ever filter up the chain of command in a format that everyone could see.

For every department and for every meeting, we started using standard templates—Microsoft Excel spreadsheets—to record data indicating progress on a variety of performance measures.

Budget numbers were the first easy wins. This was an internal discipline our city government had never lost. Most government agencies are far more adept at telling you what government costs and spends than they are at telling you what government does or how well they do it.

The first several lines of the new departmental spreadsheets for CitiStat were fairly standard across all departments. Line one was dollars spent with minority- and women-owned businesses on a year-

Bureau of Solid Waste: Moving in the Right Direction

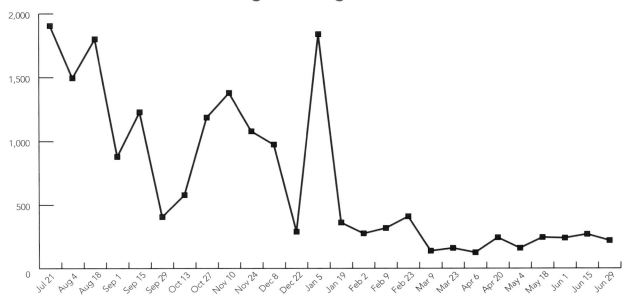

Reduction in overtime hours (above) and reduction in complaints and lost workdays (below) related to refuse collection over a one-year reporting period.

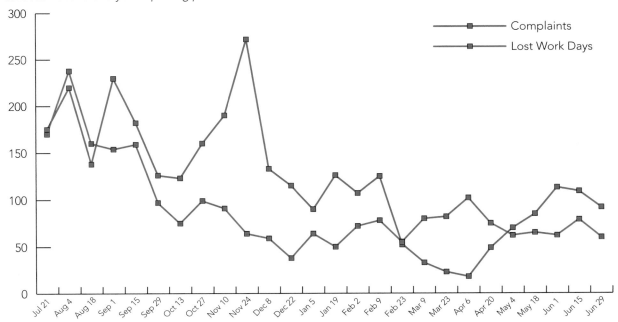

It is all connected—reduction in citizen complaints, lost workdays (unexcused absences), and overtime usage. We quickly figured out that these were the primary measures to drive for improvements in refuse collection that would make our city's streets and alleys cleaner. Note the measure in two-week reporting periods. People work against deadlines. Do you want 1 percent annual progress, or 1 percent progress every two weeks?

to-date basis. (We made it the first line intentionally because we had the most ambitious goals in the nation. We also had never before hit our goals, but soon we would.) The next lines were the amount of budgeted dollars spent on a year-to-date basis. These lines were followed by the data on unexcused work absences, and its related cousin, dollars spent on overtime year to date.

For the budget numbers, we could see the spending on a year-to-date basis and compare the rate of spending to the same date of the prior year. And this allowed everyone to see plainly whether we were over-spending or under-spending on our budgets. But for other outcome metrics, we were starting from scratch, and we would have to develop month-over-month, and year-to-date comparisons as time, experience, and the measuring of progress moved forward.

After these initial budget, procurement, and personnel metrics, we tracked the performance measures of the unique missions of the individual departments. In a collaborative way, we worked with the experienced staff of each department to tease out the best measures of effectiveness and performance. For the Health Department, it could be the number of addicted people currently in treatment or the number of children immunized against disease. For Recreation and Parks, it could be the number of children participating in recreation center activities. For Transportation, it could be potholes and curb repairs.

Occasionally at the outset, we would fall into the trap of trying to measure everything from soup to nuts. We learned to avoid such pitfalls. Soon we zeroed in on the "primary colors" of performance for any given department—those monthly indicators of performance and citizen satisfaction that together painted a clear picture of operational progress. We got better at focusing in on the leading actions that improved performance.

And whatever we measured landed on the map.

Separate departmental silos of information were all forced to land their base on the same map. Thanks to GIS technology, we could see where the problems were and where our money and efforts were landing. We could see where our problems were clustered or concentrated. We could see which teams of city workers were really performing at a high level, and which were not; which neighborhoods were being well served, and which were not. We could identify patterns and get inside the turning radius of the problem.

Who? What? When? Where? Why?

Executive Focus, Accountability, and Repeatable Routines

Every two weeks and all year round, each department head and their command staffs would meet with me and my command staff. This became our repeatable routine: the recurring wheel of accountability and executive focus. Ten departments rotated through CitiStat every two weeks. Every time one meeting ended, we knew in thirteen days we would be together again. And again, and again, and again.

Soon there wasn't much need to schedule a one-off meeting with people you knew you were going to see next week—you could grab them before or after CitiStat. What conceptually sounded like a time-waster instead turned out to be a huge time-saver.

A Consistent Presence

Did I attend every meeting? Yes, at first. And over the course of time, I personally attended fewer meetings. But I figured out ways to somehow be present at every one. Let me explain how.

The CitiStat Room was designed as a circle (actually, a semi-circle), with the department head at a podium flanked on either side by her or his command staff and the mayor facing them on the other side of the circle flanked by my command staff.

On my command staff were those cabinet members whose duties had citywide impact on the operations of every department. Their duties, like my own, were cross-functional and enterprise-wide. Some cities refer to this group as the "executive cabinet." They include the finance director, city solicitor, labor commissioner, human resources director, and head of information technology. Leading the discussion every week would be the CitiStat director and/or Michael Enright, my first deputy mayor. Depending on the subject matter being discussed, one of the other three deputy mayors would also be at the CitiStat table according to their portfolio of responsibility.

The news media had a hard time understanding the collaborative nature of what we were about. Early news stories compared CitiStat meetings to a mayoral firing squad for bad department heads. For us, it was a repeatable routine that kept us all focused on our mission. It was a precious hour every two weeks to focus on the vital functions of one department. The sharper

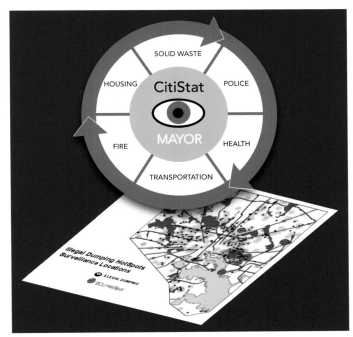

Executive wheel of accountability and focus: The repeating pattern of short collaborative meetings—focused on the latest emerging truth—allowed us to pivot or persist based on the latest evidence of whether what we were doing was working or not. What we need isn't so much new technology, but new social-technical practices, routines, and disciplines in pursuit of strategic goals.

departmental leaders soon realized it was an opportunity every two weeks for them to let me and my command staff know how we could better support them in doing their jobs more effectively.

It was a time for leaders to rise.

At every meeting, each of us had a copy of a memo laying out the major areas up for discussion. The department head would stand at a podium with the graphs, maps, and numbers measuring their department's performance. All this data was projected onto large screens behind them. And also, up on the screen was a photo of the department head, bureau chief, or manager who owned the numbers on the screen—just in case anyone was confused about which leader was responsible for what performance.

The beginning questions—on any topic, really—were always the same:
- Are we doing better over these two weeks than the two weeks before?
- If not, why not?
- And how can we improve?

This was not a firing squad. It was not a cross-examination—although, sometimes tough cross-examination was required to get to the heart of the operational challenge. It was first and foremost a conversation about how we could better collaborate to improve performance. Mapping the problems and collecting the numbers was not an end, but rather the methodology which helped us to understand how well we were serving the people we worked for.

Fiscal Year 2001 Impacts of CitiStat Initiatives

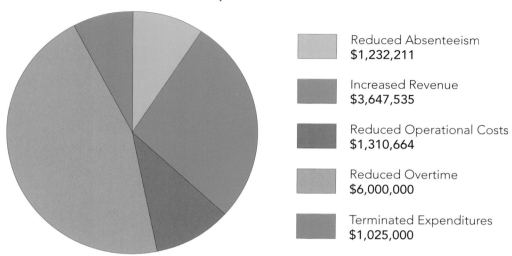

Reduced Absenteeism	$1,232,211
Increased Revenue	$3,647,535
Reduced Operational Costs	$1,310,664
Reduced Overtime	$6,000,000
Terminated Expenditures	$1,025,000

Total Estimated Impact: $13,215,410

Many wonder about the cost of starting up a performance management system. In truth, the cost was minimal—some wall board, some projector screens, a couple of computers, and a handful of committed young staffers. The cost savings in the first year alone paid for those small costs many times over.

By mapping citizen complaints, for instance, we were able share a real-time picture of what was going on in our daily efforts to address those complaints. At one meeting, a map might clearly show sanitation complaints like dirty alleys, graffiti, and illegal dumping. The accompanying graph would show the progress, or lack of progress, over the preceding weeks, months, and years. What are we doing about the spike in illegal dumping complaints in this neighborhood along this corridor? When is it happening? Who is doing it? What can we do to stop it?

At another meeting, we would be looking at a map showing complaints about cleaning and boarding up abandoned houses. This, too, would be accompanied by a graph marking our progress. And another chart showing our backlog of complaints. Is the backlog growing or is the backlog being reduced? How many of these complaints are for houses we already cleaned and boarded up before? What can be done to better secure these structures so citizens don't have to call us back, time and again?

Other meetings were about how our recreation centers were measuring up, based on the results of regular field inspections. Was playground maintenance happening? Were swing sets being repaired? Were the nets on the hoops of our basketball courts being regularly replaced? Were water fountains functioning?

Small things done well make bigger things possible.

Updated data submissions from the departments were due by noon the day before any CitiStat meeting along with any maps, charts, or graphs the department wanted to highlight. These would be supplemented by the CitiStat analyst for the department.

For every meeting, there was an agenda prepared by the responsible CitiStat analyst with suggested questions for the panelists. We typically had time to cover three or four topics after a brief update and close-out of follow-up items from the prior CitiStat meeting for that department. Department heads were given the main topics of the agenda in advance. They were not given the questions.

At every meeting, a person sat directly behind my place at the panel to take follow-up notes. Memos detailing the commitments everyone made during those meetings were typed up and distributed by the close of the same day. Everyone involved had been at the meeting, so everyone now knew what commitments had been made, and everyone knew what was expected from whom by the next meeting. If they had any doubts, they now had the memo. If there was something unclear about a follow-up commitment, it was up to the recipient of the memo to pipe up for clarity.

The meetings started on time and ended on time. Occasionally, brief huddles would happen immediately following the meetings. Sometimes, if thorny personnel matters or leadership issues had to be addressed, we would meet behind a closed door in a side room—a room that became popularly referred to as "the time-out room."

When the ramp-up of departments was complete, the CitiStat Room was a busy place week in and week out. Most meetings were held in the mornings. Sometimes, there were two or three meetings in succession. As one meeting ended, another would come in. The core team of three to five CitiStat analysts leap-frogged over each other as the biweekly rotation of CitiStat sessions progressed. The CitiStat director and our first deputy mayor generally split up responsibilities so that at least one of them was there to drive the meeting. Their backups were one of the other three deputy mayors. On occasion, an analyst might have to drive the questions and the meetings.

Everyone knew the CitiStat director and the deputy mayor spoke for me. On days when I could not be present for a whole meeting, I was focused for the part I was. Some days when I had paperwork or thank you notes to write, I would sit behind the panel against the wall—observing and listening, seen and overseeing, but not demeaning the questioning process by pretending to be fully present in the circle. Some days, I would watch just a portion of a meeting from the projection booth on my way in or out of City Hall for other appointments. On days I could not be there at all, I had the agenda for the CitiStat meetings in my slim three-ring daily schedule binder. I would write notes and questions in the margins for a job well done, or to highlight a problem that was only getting worse. And then I would make sure the agenda with my hand-written notes and questions made it to the CitiStat Room and the person driving the questions.

In these ways, I maintained a presence. An executive presence. And it was always close to the latest emerging truth—the truth that our entire government was working, in some way, to improve and change.

Recognizing Leadership

Some department heads and managers took to the new process more quickly and happily than others. The faint of heart and the change-averse could try to wait us out. Some tried. But they did so at their own professional peril because new leaders were emerging. And those new leaders would gladly take their place in a system that so clearly recognized effort, leadership, and performance.

At the beginning, we had to overcome huge training and compliance issues. In important parts of our workforce, we had to overcome not only computer literacy challenges but literacy challenges. And some managers just felt threatened by the change. "Yes, I understand that your brother-in-law wrote your special software in your department especially for you. But we are no longer using it. We want you to collect the same information every week in the open format all of us are using, so that everyone can see it on the map or on the screen."

I had promised myself that I would not wait for the whole city to tell me when a leadership change was needed in one of our departments. Some people are very capable at many things. But not everyone is good at being an effective leader in the Information Age. By the end of our first year, we were on our second police commissioner, our second commissioner of Housing, our second director of Recreation and Parks, and our second head of the Bureau of Water and Wastewater. They were all honorable, hard-working, and decent people.

311 and CitiTrack: One Call for Customer Service

One of our most important innovations—complementing CitiStat—was an idea we borrowed from the City of Chicago.

The idea was to have one phone number, with trained call-takers, for every type of service for which a citizen might call. Seems simple. Seems like a no-brainer. This is how 911 had been helping police and fire respond to service calls for decades. But in 1999, the idea of applying 911-level service to all citizen calls was new. Only Chicago, under Mayor Richard Daley, had done it.

In the old days, council members gave out jumbo cards at community meetings listing 300 different phone numbers. Trash problem? Look under "T." Dead tree? Look under "D"—and you had better have your glasses handy for the fine print.

Typically, city employees who answered such calls had not earned the assignment because of their excellent phone manners. Too often they were problem employees that could not be trusted out in the field. Managers stuck them inside where they could keep an eye on them—answering phone calls from our customers and bosses—the citizens. It should not have been surprising that calls were sometimes dropped, handled poorly, handled rudely. And we wondered why citizens were angry.

The more persistent citizen might follow up by calling each of their three council members, the council president, the mayor, and tell each of them they are their favorite. Then a week later when the pothole or dead tree issue was still unaddressed, they call their cousin who claims to know someone in city government. Or their aunt who used to date a guy at Transportation. When the problem still didn't get addressed, they would start over. Or they might start looking to move out of the city to the county where services are better.

It was no wonder that people had lost all trust in their city's ability to deliver.

We borrowed from Chicago's pioneering work and in 2001 became the second big city in America to implement a single customer service call center: Dial 311 for all city services. Along with the call center was the ability—just like 911—to monitor dispatch and delivery of service. For the first time, every citizen could now receive an individual customer service number for tracking their service request to delivery *and* a time frame (a deadline) within which to expect the service to be completed. Add to that a courteous call-taker, and it was a huge winner with citizens.

We called it CitiTrack. And the ability to independently monitor service requests and the timeliness of their delivery was a tremendously powerful pipeline of evidence that came to inform every CitiStat meeting. CitiTrack not only informed the decisions of the second new tenet of city government—rapid deployment of resources—CitiTrack also informed every tenet from timely and accurate information shared by all, to the effectiveness of our tactics and strategies, to the relentlessness of our follow-up.

Jack Maple predicted, "If you get a stat process going for all your major departments, you'll get this jalopy up to twenty-five miles per hour. But if you put a 311 number for all city services on the front end, like the police already have 911, you'll get her revving up to forty miles per hour!"

He was right.

For the first time, all complaints were on an equal footing—you didn't have to know someone at City Hall or be from a particular neighborhood to get timely service. And for the first time, a service request could be traced from call to completion. Not one was lost. And 94 percent were resolved on-time by the responsible agency.

We trained the staff of every council member's office so they could give their constituents a customer service number, get the complaint into the system, track it to completion, and get the political credit for delivering results.

In the first year, we took 656,546 calls. Most could be handled on the phone, but 221,498 resulted in service requests. We responded to more than fifty thousand snow and ice calls alone, with many city employees working twenty-four-hour shifts and taking naps on the floor during storms. We were collecting more fines for illegal dumping, and we were towing and auctioning off so many more abandoned vehicles that our impound lots were overflowing. The number of outstanding complaints about dirty lots and alleys and illegal dumping decreased steadily, from 2,717 to 619. The average time to resolve those complaints decreased from sixty-four to fifteen days for dirty alleys, from 120 to ninety-four days for dirty lots, and from thirty-one to fifteen days for illegal dumping.

CitiTrack and 311 were not magic elixirs. Nothing could instantly cure every operational deficiency, including a city work force that had over the previous decade been cut by more than 20 percent. And none of the people who had abandoned more than fourteen thousand properties in the 1990s had left an endowment for maintenance, boarding up windows, and cutting the grass.

But for the first time, we had timely and accurate information, a single source of truth, a common platform for solving problems and more prudently deploying our limited resources. Most important, we had a new level of accountability that was beginning to permeate every level of city government.

Tackling Trash

One of our more visible, overwhelming, and socially demoralizing problems in Baltimore was the illegal dumping of trash. Within three years of establishing CitiTrack, we were addressing our dumping complaints in weeks rather than months. And, we managed to identify and apprehend a tiny number of unscrupulous haulers who were making life miserable in some of our poorest neighborhoods.

Most elected officials spend the day putting out fires, reacting to the most recent crisis. But Jack Maple's third tenet—effective tactics and strategies—had us focusing on long-term solutions. When we walked into City Hall, there were thirteen trash collection boroughs, just as there had been since the early 1960s when Baltimore had many more residents. Trash pick-up routes were not based on current needs. Some crews finished in a few hours, while others went into overtime every day. And borough supervisors were responsible only for trash collection, not for keeping their districts clean.

We changed tactics by using GIS technology to reconfigure our districts, equalizing the workload and creating a level playing field so we could compare how different crews performed. As a result, we were able to identify and showcase the real leaders. Then we changed strategies by making borough supervisors responsible for keeping their districts clean—trash and bulk trash collection, recycling, and alley and street cleaning.

We ran a contest. There were cash prizes for boroughs that did the best job as measured by citizen complaints, absenteeism, and tonnage collected. A crew that started near the bottom ended up winning first place (and most of the others improved). The leader of this borough was quickly promoted. The crew that started in last place stayed in last place. When we did a deeper dive on the data, we saw that the crew made more money by rotating unexcused absences (thereby juicing their overtime pay)

311 Dashboards Track the Status of

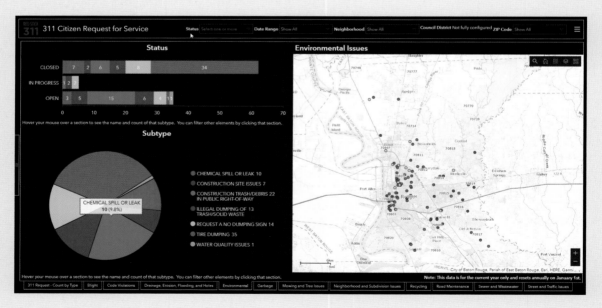

The public can use the City of Baton Rouge, Louisiana's 311 Citizen Requests Dashboard to view 311 requests across the city. The filters along the top banner can be used to filter by status, date range,

Citizen Complaints, Requests for Service

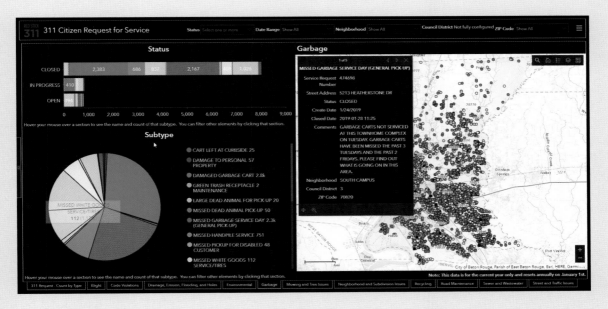

neighborhood, council district, or zip code. The tabs at the bottom can then be used to view a detailed map and additional info about each request type (environmental issues, garbage, road maintenance, etc.).

than they would have made chasing one of the prizes. The leader of that crew was soon replaced.

Cash prizes are hard to craft and manage. Other things are easier, like thank you notes. I sent personal thank you notes to high-performing work crews. And I never felt like I did it enough. Or a pair of tickets for the mayor's seats to see the Baltimore Orioles, or in the mayor's box to watch a Baltimore Ravens game.

Early in our administration, one of the managing partners of a major downtown law firm called a friend of mine to lodge a delicate complaint. It seemed that when this managing partner brought high-paying clients to his firm's sky box for a Ravens game, the guests in the mayor's box next to his "looked like garbage collectors." My knowledgeable friend told the complainant they were not just any garbage collectors; they were the best garbage collectors in the city, and they had earned their seats.

Filling in Potholes

Some naysayers publicly accused our new crime-obsessed administration of having absolutely no vision beyond crime. So, we responded with a forty-eight-hour pothole guarantee.

In just the first few weeks of CitiStat, we reduced the delivery time on filling potholes to fifty-three hours. Hitting forty-eight would be a stretch, but we were confident—even with an uptick in volume—we could hit it.

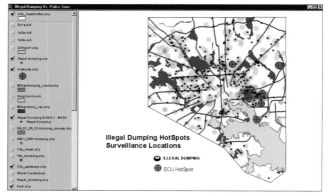

A hot-spot map of illegal dumping from a CitiStat meeting in 2001. If you only had two inspectors who could look for illegal dumping throughout an entire city, where would you have them look? If you had a limited budget for placing cameras that could captured the culprits and their license plate tag numbers, where would you place the cameras?

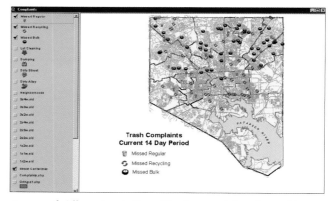

A map of different varieties of trash complaints from citizens in one overlay from a CitiStat meeting in 2001.

We held a press conference to publicly announce the guarantee. A worried Department of Public Works spokesperson had prepared a big sign announcing a "48 to 72 Hour Pothole Guarantee." I wasn't sure if he was trying to protect his co-workers or me, but I commandeered a big black Magic Marker and emphatically crossed out "to 72"—leaving only the bold promise of 48 hours as the news cameras rolled.

The television reporters were very excited to see us fail—honestly, to see me fail. They fanned out across the city and called in complaints for potholes just to set up on them with their cameras and watch them fail to be filled in time. They were greatly disappointed.

We had to burn some overtime hours at first, but our crews rose to the challenge. They took pride in doing their jobs well. And within two weeks, they hit the goal 98 percent of the time.

Neighbors shared their experiences. A timely pothole response! Astounding! People calling from poor neighborhoods received the same guarantee as people calling from wealthier neighborhoods.

Democracy in America was at work. The newspaper covered the announcement of the guarantee, but not the fact that we hit the goal. It didn't matter—soon the whole town knew by word of mouth. The doubting among us tried it out for ourselves: "I'll be damned, they actually came in forty-eight hours and fixed it."

For the first time in memory, every worker on our pothole crews received a personal thank you letter from the mayor. Thank you notes to city employees are the most underutilized tool in most mayor's offices.

We acknowledged the job well-done. We lifted up the leaders.

Dealing with Vacant Houses

Potholes are one thing, and one department can address them. But many problems are more complex. They require effective action from more than one department. We started running plays—synchronizing the activities of two, three, or four different departments to achieve a city objective.

It is still difficult for any visitor from another city to drive through parts of Baltimore or look out the window of a passing Amtrak train without being disturbed, at a gut level, by the specter of abandonment. The vacant, dilapidated houses were left behind by the biggest population loss of any city in America. This is not how any part of our country is supposed to look.

At its peak in the 1950s, Baltimore housed approximately a million people. By 1999, we were

410-POT-HOLE

NO IFS, ANDS OR BUMPS!

Repairs will be made within 48 ~~OR 72~~ hours.

Guaranteed!

Mayor Martin O'Malley

Our original "48 to 72 Hour Pothole Guarantee" poster—note how my own cautious staff tried to protect me from the political vulnerability that comes from setting an ambitious goal. I crossed out the 72 when—contrary to prior directive— it appeared in the big sign at the press conference. The assembled media were salivating to report on the failure, but our crews rose to the challenge and delivered.

down to about 630,000. Drug violence and addiction had not only claimed a lot of lives; it also robbed thousands of families of that building block of legacy wealth and economic security we call a home. Elderly widows whose husbands had won the Second World War became prisoners in the homes where they'd raised their families—as their neighborhoods deteriorated and drug violence spread. When these widows died, the cost of probating the property was often more than it was worth. We ended up with more than fourteen thousand abandoned houses, at one point, which had become shooting galleries for junkies and prostitutes, flop houses for the mentally ill and the homeless, and clandestine arenas for staging pit bull fights. They were dumping sites for trash, breeding grounds for rats, in-your-face sources of stench and demoralization for every neighbor on the block, and a health hazard for anyone unlucky enough to live next door.

During the first of what would become our twice-a-year cleanup weekends, I walked into an alley where the city's front-end loaders and dump trucks were rolling in to clear huge mounds of trash representing years of illegal dumping. A young mom came running down the back steps of her home and rushed up with tears in her eyes, holding her diapered infant up to me in gratitude. I had seen newsreels of American troops being greeted by young moms proffering their babies in the same way in French villages liberated by the Allies during World War II.

What a shame that we had allowed ourselves to forget so many of our people. Trash, rats, abandoned buildings—these are not small nuisances if you are raising your infant child in their midst.

As we ramped up HousingStat, we counted a backlog of six thousand trashed row homes that needed to be addressed—and these were just the ones the neighbors hadn't given up calling to complain about. The problem had never been handled systematically. It was first come, first served, and "we'll get to your complaint in eight months or so." Maybe.

A map of vacant houses from a CitiStat meeting in 2001. Baltimore suffered more population loss from 1969 to 1999 than any other major city in America. When people fled neighborhoods where housing values had been decimated by violent crime and open-air drug dealing, 15,000 vacant and abandoned homes were left behind.

On a scale of one to ten in municipal difficulty, this was a three. We started by mapping the complaints. We recorded the dates they came in and the number of days it took to address them. And we started tracking—not only the address of every nuisance, but the number of different calls made by different neighbors about each one.

A picture emerged.

We started asking questions about how the work orders were being handled. Were whole blocks ever batched together to be knocked out at once? No, they were not. Crews simply went down their lists and drove all over the place to clean or board up whatever house popped up next.

After a crew from the Bureau of Solid Waste cleaned a house, how long did it take for a crew from the Housing Department to show up with tools and plywood to board up broken doors and windows? The answer was at least several days, but it could take weeks or even months.

We learned that when a Housing crew did arrive, the vacant house had often been trashed again, and the Housing crew couldn't board it up until Solid Waste returned to clean it again.

We learned that when Housing crews finished their work, the people who used the vacant homes for drugs or prostitution would usually return that very night, pull the boards off, and resume occupancy.

We brought crew leaders from both agencies—Solid Waste and Housing—to meet in one room. After acknowledging the way we had always done things, we asked them to come up with a better way.

We merged crews from the two departments and created new crews that would have the ability to both clean and board. We carved up the map into smaller work areas. We dedicated half of our crews to the most concentrated areas of complaints. For these crews, there would be no more running around. They would work on the same block until all complaints were resolved and jobs completed. Other crews would handle geographically scattered complaints.

We began to overlay our 311 housing and sanitation complaints with 911 complaints about drug dealing and prostitution. In the worst areas, instead of boarding houses up with hammers and nails, we

Governing for Results: CitiStat

In just the first year of CitiStat, we achieved some compelling results.

- Tracking chronically tardy and absent employees, we cut unscheduled leave in the Department of Public Works by as much as 67 percent, meaning that more than 100 additional employees were on the job every day.
- Identifying the highest earners of overtime in General Services, Transportation, and Water and Wastewater, we saved $1.2 million in just the first five months.
- Working with building security guards and their union, we saved $400,000 by simply changing schedules—and avoided layoffs.
- By prioritizing a huge backlog of work orders in the Bureau of Water and Wastewater, we found two thousand water meters that hadn't been paying for service for years.
- We identified employees with "take-home car" privileges and cut back on take-home cars and reduced the city's fleet, saving $2 million in just the first year.
- We analyzed and measured response times for fire and ambulance services and worked through the difficult decision to close five firehouses and, with them, seven companies. (That one hurt.)
- We never stopped looking for ways to eliminate or downsize anything that did not contribute to our overarching goal: a safer, cleaner city that would be more welcoming to private investment that would, in turn, lift our neighborhoods and employ our people. A better place for our kids to grow up.
- Twelve years of population loss leveled off as the city began to grow again.

We never made perfect. We never made our city immune to future setbacks and failures. We were not able to control for the leadership and lack of leadership that sometimes followed ours.

But we did make better. And the strength of our example helped a lot of cities figure out how to do better by their own citizens.

Effectiveness, efficiency, and performance management were not words that anyone might have used to describe city government before. But with CitiStat and 311, responsiveness, accountability, and delivering results became the rule rather than the exception. Over the course of six difficult years, customer service emerged as a new guiding ethic of public service in a government "of the people, by the people, and for the people."

started using power drills and screws. When drug dealers broke the plywood, we returned with cinder block bricks to secure ground floor windows and doors.

We began to see a healthy competition among crews. We acknowledged and thanked the leaders. They started learning better tactics from one another. Neighbors started to notice a difference.

The result: More of the vacant houses we cleaned and boarded stayed cleaned and boarded. The backlog was slowly driven down. Response time improved dramatically—from eight months to an average of just fourteen days.

It would be a while before these shells would be occupied again by humans, but at least they would no longer be occupied by trash and rats.

The Legacy of CitiStat

If you let up, you lose. We took to heart Jack Maple's fourth tenet of relentless follow-up and assessment. We relentlessly tracked indicators of our progress. We pushed senior managers to leave their offices and make unannounced visits to their troops. One visit turned up barbecue grills in a repair shop alongside the gas and oil cans. Another revealed a wall-sized collage of pornography a few feet away from a poster reminding employees of the city's sexual harassment policy.

There is no substitute for command presence at the front line.

CitiStat not only improved performance, it also saved money. In the first four years, we saved taxpayers $160 million in reduced waste and improved efficiency. Neal Peirce, a syndicated columnist with *The Washington Post*, wrote that CitiStat "may represent the most significant local government management innovation of this decade." When he was the mayor of San Francisco, Gavin Newsom called CitiStat and 311 not just evolutionary, but revolutionary.

We had replaced the old patronage politics, long familiar in most American cities, with a meritocracy. And a collaborative meritocracy at that. CitiStat raised expectations. Residents saw improvements in their neighborhoods, but they were also able, for the first time, to see for themselves what we were doing. They could check weekly CitiStat reports on our website and read about our progress—or lack of

Beyond Baltimore

As CitiStat gained traction in Baltimore, an interesting thing happened: officials from every level of government from across the country and even as far as Europe and Asia—be it from townships, cities, states, and even federal agencies—began sending delegations to Baltimore to learn more about CitiStat. Their interest has not been surprising. Whether you work in the public, private, or nonprofit sectors, being able to regularly measure and evaluate performance is critical to successfully managing any endeavor. Whether you're responsible for managing a small unit of less than five people or a large city with sixteen thousand employees and more than six hundred thousand residents, a program like CitiStat can provide the critical information you need for better decision-making, and be the driving force behind your most important management and service initiatives. It can provide the structured mechanism to ensure that you stay focused on your top priorities.

progress—in a weekly email newsletter. They could see the same maps of service requests and service delivery that we were using to track and measure performance at City Hall. CitiStat began to restore the trust we had lost in one another by giving us the responsive government we all needed and desired.

If you Google "CitiStat," you'll now see it popping up across the country. Growing numbers of county commissioners and executives have initiated their own CountyStat performance measurement systems. Which is the best testament to any good idea—that other people want to adopt it. It's what we did in adapting the CompStat process from New York and borrowing 311 from Chicago.

But CitiStat is also about leadership. It requires leaders who are willing to demand and closely monitor progress. It requires leaders who are willing to be vulnerable and accountable to the public. It requires a suspension of fear. It requires a willingness to declare goals and let everyone follow the ups and downs—and there will be downs—of your performance. There were many mayors from other cities—mayors with maybe six or seven years of experience under their belts—who walked into the CitiStat Room in Baltimore, saw how openly we were doing what we were doing, then turned to their able staff and whispered, "Let's get the hell out of here. Do not tell anybody we saw this."

But the newer ones, who didn't have the burden of explaining past failures to perform, tended to embrace it—mayors like Adrian Fenty in Washington, DC; Manny Diaz in Miami; Matt Driscoll in Syracuse; Gavin Newsom in San Francisco; Byron Brown in Buffalo; Cory Booker of Newark; Michael Bloomberg in New York City; Joe Curtatone of Somerville, Massachusetts; and countless others.

After our administration moved onto Annapolis to run state government, Baltimore's CitiStat had its ups and downs. At first, it continued with the new energy of my successor's administration. Several patriotic public servants stayed in the city, at the new mayor's request, to guarantee continuity. But mayors change, and by 2013, CitiStat had fallen into disuse. Meetings stopped happening regularly. Then they stopped happening at all. When the press finally noticed and wrote about it, the political embarrassment and declining levels of city service resuscitated it for a time. Now, yet another mayor has vowed to reinvigorate CitiStat—she has rebranded it "CitiStatSMART." I hope she succeeds. It will require humility, discipline, and relentless follow-up.

Time will tell, and so will the citizens. This openness and transparency stuff can be kind of scary for politicians, but it works best for the people.

Learn & Explore

Los Angeles Clean Streets Index
As part of its Clean Streets initiative, the Bureau of Sanitation quarterly drives all public streets and alleys and gives each block a "cleanliness score" on an interactive map.

Blight Status Dashboard
This app can be used by local government leaders to proactively monitor the status of blight complaints and efforts made to reduce blighted properties.

For links to these and other examples, exercises, and resources, visit SmarterGovernment.com and click chapter 6.

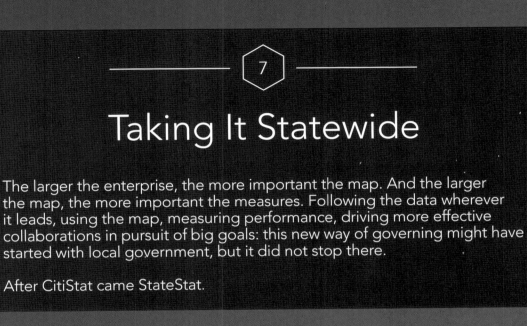

7

Taking It Statewide

The larger the enterprise, the more important the map. And the larger the map, the more important the measures. Following the data wherever it leads, using the map, measuring performance, driving more effective collaborations in pursuit of big goals: this new way of governing might have started with local government, but it did not stop there.

After CitiStat came StateStat.

Landing the Effort on the Map's Targets

This new way of governing is not just for cities—it's also for states. True story.

On August 1, 2007, during evening rush hour, the I-35W Mississippi River bridge in Minneapolis, Minnesota, collapsed. Thirteen people were killed and 145 were injured. Maintaining that bridge was the responsibility of the State of Minnesota, with partial funding from the federal government.

Our nation watched the horror as moms and dads, sons and daughters, plunged to their deaths due to the collapse of a bridge that everyone knew, or should have known, was structurally deficient and in urgent need of repair.

Shortly after this tragedy, I introduced Jack Dangermond of Esri to Governor Rendell. Jack had a bridge app Ed needed to see.

Jack knew that Governor Ed Rendell of Pennsylvania had a passion for infrastructure and the proper funding of America's woefully underfunded transportation needs—including bridge repairs. He had no doubt seen Governor Rendell speak out passionately in the wake of this tragedy for better and timelier investments in our nation's failing infrastructure.

But what Jack did not know was that Governor Rendell, like most former mayors, is blessed with both an aversion to blind meetings and a very, very short attention span.

When I finally succeeded in corralling a busy and skeptical Governor Rendell into a corner with Jack's computer screen, I warned Jack, "You have forty-five seconds."

Jack launched immediately. "This," he said, "is a map of Minnesota, and each dot on this map shows where the bridges are in Minnesota."

Click.

"Here's a red, orange, yellow, and green color code on those dots now showing the range from the most structurally sound to the most structurally deficient."

Click.

Then he said, "Now I'm going to click on this next layer, which will change the size of these dots relative to how many human lives go across these most structurally deficient bridges every single day."

Click.

Finally, he said, "I want to show you where federal repair dollars for these bridges are actually going..."

And with that, he clicked the final key—and the dollars scattered all over the place.

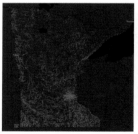   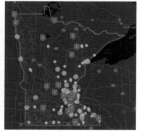

All Bridges   Bridge Conditions   Traffic Volume   Bridge Investments

The maps of bridges in Minnesota that Jack Dangermond shared with Governor Ed Rendell of Pennsylvania. Green dots are bridges. The color-coded red-green dots reflect the structural integrity of the same bridges. The size of the dots show how many human beings use the particular bridge every day. The green dollar signs show how few of the federal dollars for bridge repair and maintenance actually land on the biggest targets. But they all land on the map.

It took about two seconds for Governor Rendell to exclaim, "None of the dollars are landing on the targets!"

Dangermond immediately replied, "Not yet...but they are all landing on the map."

Sometimes the most important truths can also be the most obvious—if we know how to look for them.

Our job is to make sure we land the effort on the targets.

Applying CitiStat Techniques across Maryland

During my first couple of years as mayor, the City of Baltimore had the support of a Democratic president and a Democratic governor. During my final few years as mayor, we no longer had help from either office. Nascent collaborations between the city and the state that had started under the former Democratic governor began to evaporate quickly under the newly elected Republican governor.

Having stayed in the city rather than running for governor in 2002, I was now regarded as his next most likely Democratic challenger.

The adversarial political relationship wasn't good for the city. And it wasn't good for the state. One of the two of us had to go. And when I ran for governor in 2006, the people decided it would be the other guy.

In those days before our country was rocked by the Great Recession, we campaigned on a broad, ten-point plan to move our state forward.

Maryland Inauguration Day, January 19, 2011, after re-election. Left to right: William (14), Jack (8), and Katie.

And our plan went like this:

"Number one, we are going to make our government work again.

"Number two, we are going to make our government work again.

"Number three, we are going to make our government work again..."

The knit eyebrows in the crowds soon turned to smiles, then to laughter, and then to applause as the point of the repetition dawned on them. The list went on to include making our state safer, improving education, improving the health and well-being of our people, and restoring the health of the Chesapeake Bay. But the larger contrast with my opponent was made by priorities one, two, and three—to make progress on all these fronts, we needed to make our government work again.

After a hard-fought campaign, we won by a seven-point margin. We were the only challenger in America that year to defeat an incumbent Republican governor. And our largest margins of victory came from the poorest neighborhoods of Baltimore—the people who had seen, most clearly, how even our most troubled places can be improved when we actually make our government work again.

My successor as mayor pledged to keep CitiStat going in Maryland and to build upon its success. And she did.

As governor-elect, I promised we would bring this new way of governing to the state government of Maryland. And so we did. We called it "StateStat."

Within ninety days, the new StateStat office was built-out, complete with a separate projector

What Is StateStat?

StateStat began in Maryland in 2007 with a few select public safety and human services agencies. By 2015, StateStat was the core management component of thirteen individual agencies and hosted a number of cross-agency stat systems, such as the award-winning BayStat (see chapter 10).

Most governments monitor their performance at annual budget reviews—if they track performance at all. With StateStat, we monitored agency performance monthly, and in some cases bimonthly, identifying data trends before they turned into problems. Through relentless follow-up with our agencies, we ensured that the solutions we crafted together were not only implemented efficiently and quickly but were effective in turning the data trends back in the right direction.

Our StateStat process—modeled very much on the CitiStat process—consisted of several repeatable routines for convening and focusing the attention of the group on the most leading actions that drive us to achieve strategic goals:

• Before the meetings: agencies submitted customized data templates each month. The StateStat team analyzed the data to identify trends, conducted site visits, and met with agency staff to evaluate programs. The analysts turned this analysis into detailed executive briefing memos which were shared with the StateStat Panel, including the governor, prior to each meeting.

booth and large screens permanently mounted on the walls. Shortly after its completion, every major department began to rotate through the new performance-measured regimen. A strong, pre-existing Esri basemap provided the common platform. New cabinet secretaries provided the new leadership, and we began our experiment anew—to see if this new way of governing could be accomplished at the more complex level of an entire state government.

No plan survives the first engagement. The state was larger than the city. The bureaucracy was more complex. The enterprise responded more slowly to directives. We had to adapt.

Over the course of time and experience, a few adaptations emerged to distinguish StateStat from CitiStat. The most important of these adaptations were the result of conscious and intentional changes. Among them—the necessity of setting strategic goals, the concept and disciplines of delivery, the importance of cross-functional collaborations, and the imperative of greater openness to innovation.

A few words on each adaptation appear later in this chapter.

The Difference between Cities and States

Our mission statement as a new state administration was:
1. To strengthen and grow the ranks of a more inclusive, diverse, and upwardly mobile middle class.
2. To improve public safety and public education in every part of our state.
3. To expand opportunity—the opportunity to learn and to earn, and the opportunity to enjoy the

- StateStat meetings: the director of StateStat leads the StateStat Panel, which included the governor and/or lieutenant governor, the governor's chief of staff, the governor's legal counsel, and staff from the Departments of Information Technology and Budget and Management. The panel questioned agency leaders on the trends identified in the executive briefing memos and worked with these leaders to develop solutions. Agencies brought a variety of staff to the table, including their secretary and deputy secretaries, human resources, finance, and program staff to assist in the discussion. StateStat meetings were innately collaborative—not only did the panel ask questions of the agency but the agency could use the time to ask for assistance or guidance from the governor, his senior staff, legal counsel, Information Technology, and so on.
- After the meetings: the StateStat analysts prepared comprehensive follow-up memos for the agencies detailing the action items discussed in the meeting, as well as any other questions or concerns. The agencies completed and submitted the follow-up memos prior to the next StateStat meeting. The analysts worked continuously with their agencies throughout the month in between meetings to ensure progress was being made quickly and efficiently.

And to support our goals of transparency and open government, the StateStat team posted comprehensive meeting summaries online, including all data and charts used in our analysis. We kept an executive dashboard updated and prominently displayed on our website so citizens could see the measures I was watching. This allowed citizens to view and interact with the data anytime through our open data portal and track the progress we were making toward the administration's sixteen strategic goals.

Implementation Timeline
for Maryland's StateStat

2007 to 2014

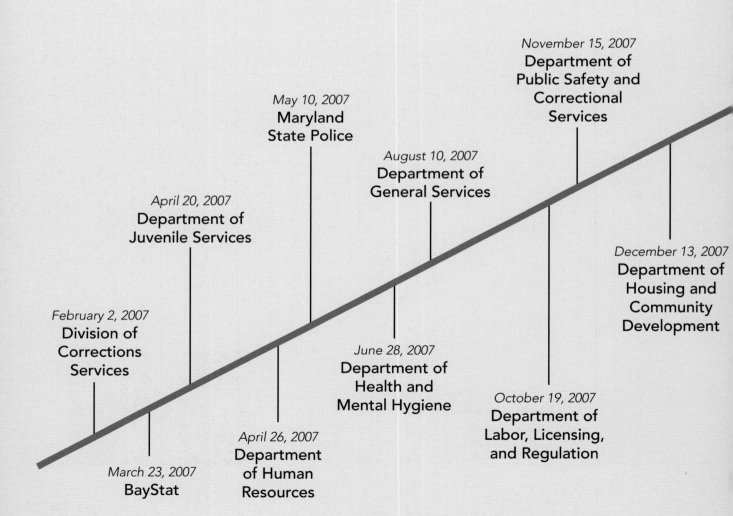

November 15, 2007
**Department of
Public Safety and
Correctional
Services**

May 10, 2007
**Maryland
State Police**

August 10, 2007
**Department of
General Services**

April 20, 2007
**Department of
Juvenile Services**

December 13, 2007
**Department of
Housing and
Community
Development**

February 2, 2007
**Division of
Corrections
Services**

June 28, 2007
**Department of
Health and
Mental Hygiene**

October 19, 2007
**Department of
Labor, Licensing,
and Regulation**

March 23, 2007
BayStat

April 26, 2007
**Department
of Human
Resources**

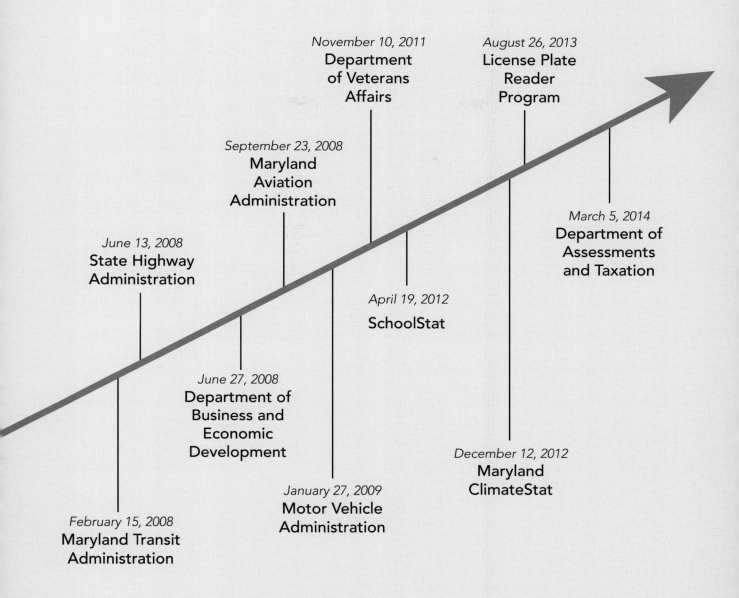

November 10, 2011
**Department
of Veterans
Affairs**

August 26, 2013
**License Plate
Reader
Program**

September 23, 2008
**Maryland
Aviation
Administration**

June 13, 2008
**State Highway
Administration**

March 5, 2014
**Department of
Assessments
and Taxation**

April 19, 2012
SchoolStat

June 27, 2008
**Department of
Business and
Economic
Development**

January 27, 2009
**Motor Vehicle
Administration**

December 12, 2012
**Maryland
ClimateStat**

February 15, 2008
**Maryland Transit
Administration**

One department and one initiative at a time, we ramped up to cover the entire enterprise of state government. The cadence of meetings shifted from CitiStat's "every two weeks," to more of a monthly cadence with StateStat. We started and we did not stop.

STATE STAT
MARYLAND

health of the people we love and the natural places we need. And to make those opportunities available to more of our people, not less, including future generations.

It was a good mission statement. It was also a good statement of governing philosophy and intent. With some tweaking, it might also be a good book of proverbs. But what did it mean in terms of actions, priorities, and choices? What were the strategic goals and the actions necessary to achieve the larger mission?

Now that I was governor, suddenly measuring the performance of city government looked a lot simpler than measuring the performance of state government.

People have sometimes asked me, "Which job did you enjoy most—mayor or governor?" The answer is "Mayor; definitely mayor." When one is effective as a mayor, you never have to explain how or why to the public. Either the alleys are cleaner, or they are dirtier. Either the graffiti vandals are winning, or the city is. Either there are no guys on my grandmother's corner slinging drugs, or there are twenty.

During my years of city service, our mission statement was routinely projected as slides on the big

The Daily Cadence of StateStat

With StateStat, every day I received one, two, and sometimes three executive briefing memos for the departments or agencies that were scheduled that day for their StateStat sessions. From time to time on a rotating basis, I would also sit in on many of these meetings, depending on the degrees of difficulty and challenge being experienced in any given area of operations.

In addition to the performance memos, I received every week a report on concerns coming into our Office of Correspondence and Constituent Services. This report consisted not only of a dashboard on the nature and proportion of incoming concerns, but copies of several representative or unique letters, as well.

I made a regular practice of always reading these memos and reports. I marked them up with questions, concerns, and where appropriate, compliments. I returned them for follow-up and response—oftentimes copying individual members of my command staff if their follow-up was required.

Reading these memos and letters, knowing whether the graphs were moving in the right direction or the wrong direction, making sure the department heads knew you read them—all these practices helped to keep me connected, present, and aware. Effective collaboration requires this sort of constant presence in the center of the circle, even when you are not physically present.

Situational awareness, knowledge, and a constantly communicated intention is what it is all about. Staying up on these memos and reports on a daily basis kept me in the center of our government. And this practice kept us focused on our strategic goals regardless of how hard the whirlwind was blowing outside.

screens before the start of every CitiStat meeting. Most city leaders around the world would likely subscribe to the same mission—"to make our city a safer, cleaner, healthier place; a better place for kids to grow up; a place where businesses want to grow and expand." The names of the departments pretty well summarized the leading actions necessary to realize the mission—Police, Fire, Solid Waste, Housing, Health, Education.

CitiStat, the 311 call center, and service request numbers made it easier than ever for citizens and city managers alike to track the delivery of very visible services. And of course, we worked every day to improve response times by fire and police, to shorten the time it took to address a service request for illegal dumping or graffiti, or to reduce big backlogs for towing abandoned vehicles or boarding up vacant houses. But everyone knew the top-line goals—a "safer, cleaner, healthier city."

At the state level, it seems so much of the work is middle-level process. And every public good is at least a four-cornered bank shot. The billiard balls sometimes make it around the table and into the right pockets, sometimes not. Visible differences to most citizens are far fewer. Only the most egregious failings make the paper. It seems every public good requires a long chain of actions—sometimes by many different departments. Many of those actions require partnerships and collaborations with fairly independent actors. And sometimes even when things go well—as with public schools or public safety— citizens naturally enough are inclined to give credit to their local government rather than their state government for the progress.

Moms and dads are proud to send their kids to a good city school, or a good county school—not to a 65 percent state-funded school. Most citizens attending a ribbon-cutting ceremony for a new university building in our state, might reasonably assume the guy in their community who bought the naming rights was responsible for 95 percent of the building costs rather than their own state tax-dollars.

Understanding what our states should and should not be doing is not just a challenge for citizens. It is a challenge for state bureaucracies and the people charged with managing and leading them. There are a few exceptions to the often invisible, mid-level work of our states—wait times for a driver's license renewal at the Motor Vehicle Administration, certain state permits, or child support enforcement—but, for the most part, very few states set very clear goals...for anything.

That's just the way it has always been. But things are changing.

Setting Strategic Goals

It took the better part of our first year in office to settle in on a clear articulation of strategic goals with deadlines. (And ours was a busy first year that included my calling a high-wire, high-stakes special legislative session to tackle a ballooning structural deficit. It was not a fun experience. We succeeded, and I still bear the scars.)

In our search for the right strategic goals, we found that the simplest goals to communicate weren't necessarily the best goals when it came to motivate, organize, or create collaborations within a big bureaucracy. The easiest goals to attain were usually tactical rather strategic. But the new StateStat regimen, and the honest, collaborative conversations with cabinet secretaries and their leadership teams, soon resulted in a framework of strategic goals. And these goals became our roadmap for eight years of nation-leading progress.

We would go on to make our state's public school system number one in the nation for five years in a row. We would reduce crime to a thirty-nine-year low. We would reduce our incarceration rate to a twenty-year low. We would recover jobs lost in the recession at a faster rate than any of our neighboring states. We would make our state number one for women in business. We would restore the health of the Chesapeake Bay's waters to their best levels since 1985. We would defend, through a recession, one of the few triple A bond ratings, as well as the highest median income in America. We would make college more affordable rather than more expensive. And the US Chamber of Commerce—which rarely says anything kind about any Democratic governor—named our state number one for innovation and entrepreneurship for each of my three final years in office.

We didn't know it in 2007, but our nation was about to head into the deepest recession since the Great Depression. And in hindsight, I can now say that a big part of what got our state to the other side of that recession before other states—and, in many cases, stronger than we were before—was the clarity of our strategic goals and the dashboards that monitored our efforts, our progress, and our results (good and bad) in ways that all could see.

It might seem counterintuitive to pursue higher goals in the face of a recession, but it worked for us—even with the brutal series of deep budget cuts that we, along with every other state, were forced to make in the ensuing months.

We no doubt violated a management maxim or two by having so many strategic goals—sixteen in all. But each of them was carefully chosen to reinforce the others, or to address a serious shortfall—unique perhaps to Maryland—that was holding back the potential of our state.

Initially, our list of strategic goals numbered fifteen. But when the recession hit, job creation became goal number one, as we realized—or perhaps remembered—there is no progress without jobs. The other goals fell into the broader categories of opportunity, security, sustainability, and health.

Attached to each strategic goal was a deadline. Each deadline had interim benchmarks. And the vast majority of those deadlines—and all the benchmarks—came due within our term of office.

We created an executive dashboard link on the main page of the state website that, in a simple, color-coded glance, showed every citizen which of these public pursuits was on track and which ones were lagging. In a clear red-light-to-green-light spectrum of up or down arrows, every citizen could see what their governor saw as we drove the leading actions of progress-making.

With a click on any strategic goal—jobs, crime, education, infant mortality—every citizen could review in deeper detail the maps that showed where we were hitting our goals and where we were not. Every citizen and stakeholder could see the graphs of leading actions and results. Everyone could see whether a particular tactic or strategy was moving us in the right direction or the wrong direction. And every citizen could read, or download to read, the written plan for achieving a particular goal—which brings me to the concept of *delivery*.

Delivery and Dashboards

I first heard the concept of delivery explained at a conference in 2007 by Michael Barber. Barber had created and headed the "Delivery Unit" of Prime Minister Tony Blair's government in the United Kingdom.

Delivery is a concept of public management that, very much like CitiStat, involves a dedicated team, use of direct reporting to the chief executive, and measures applied to drive performance in a relentless cadence of "stocktake" meetings. These meetings are frequently attended by the chief executive—in this case, the prime minister.

Unlike CitiStat, delivery emerged at a national level rather than up from the local level. And unlike CitiStat, delivery subscribes to the indispensable necessity of having a "delivery plan"—a plan of leading actions, responsibilities, and deadlines. And within these plans, graphs show trajectories of anticipated progress against which to measure actual effort and actual results. In other words—and quite

Governing for Results: StateStat

When the O'Malley-Brown Administration came into office, Maryland was facing a $1.7 billion structural deficit, too many schools were underperforming, crime was rising outside of Baltimore, and regional competitors were outpacing Maryland in the rate of new job creation.

Seven years later—through recession and hard times—Maryland had come to reap the better results that flow from making better choices. With strategic goals, a performance-measured system of government called StateStat, and a lot of tough decisions, a stronger Maryland began to emerge.

Crime was driven down to thirty-nine-year lows. Maryland achieved the fastest rate of new job creation in the region. Maryland was one of only nine states to maintain a triple-A bond rating through the Great Recession. At the same time, a series of tough fiscal decisions eliminated the structural deficit. And a truly progressive income tax was put in place for the first time.

Notwithstanding $8 billion in cuts—the largest of any administration in history—the O'Malley-Brown Administration made record investments in education, innovation, and infrastructure. Major strides were made to improve the health of the Chesapeake Bay and the health of the people of Maryland.

As the O'Malley-Brown years came to a close, Maryland had become one of the best states in America for upward economic mobility. The US Chamber of Commerce named Maryland the top state in America for innovation and entrepreneurship for three years in a row, and *Education Week* magazine named Maryland's public schools the number one school system in the nation for five years in a row.

More trees were planted; more streams were buffered; more water treatment plants were upgraded; more cover crops were planted; and more farmland, forest, and open space were protected than ever before. More renewable energy was placed on the grid than ever before. The oyster and the crab populations began to rebound in response to new sanctuaries and more sustainable practices.

Greater efforts were made to shelter the homeless, to eradicate hunger, to care for the sick. Infant mortality was driven down to a record low. Fewer children went hungry, fewer children were victims of violent crime, and more children received health care than ever before.

Better choices, better results.

Key Lessons of Delivery in Government

We are getting better at understanding and delivering success all the time. The difference between success and failure is all in the detail.

Don't start a delivery unit unless you really mean it.

But if you do mean it, here are the crucial details to pay attention to.

Leadership

Setting up a delivery unit will not happen without the right leadership at two levels. The political leader (this might be a president, a prime minister, or at the provincial or state level, a premier or chief minister or governor) must be ambitious, focused, and disciplined. Meanwhile, the head of the delivery unit needs to believe in the mission, love data and graphs and maps, be loyal to the boss, and be excellent at building relationships with politicians and officials.

The test for a delivery unit head is whether you can deliver a strong, critical message to a senior politician or official in a meeting and leave the room with the relationship stronger than when you went in. This requires a combination of courage, humility, and empathy.

Numbers

Second, move the numbers! You have to have a goal and a plan, and you have to collect data to know what is working and what isn't. That means regular flows of data as close to real time as possible.

People

Third, if the delivery unit is to be the key to driving the government's agenda, it needs to be staffed by great people. The leadership we have already mentioned, but what about the rest of the team? As the founder of Peru's delivery unit, Ernesto Balarezo said, "One of my main functions is to find, develop, and empower the best talent. If I am successful in that, all else will follow." Ernesto brought in people from outside and mixed them with people who knew and understood government.

Culture

Fourth, a successful delivery unit requires a special kind of culture. Jenny Cargill, who has led the delivery unit in South Africa's Western Cape, says it requires a team characterized by "an untiring work ethic, persistence, resilience, plain speaking, self-motivation, taking initiative, and flexibility." The environment the team works in, therefore, needs to "support innovation, constant inquiry, and an orientation towards finding solutions." By contrast, delivery units that don't succeed are always telling you how difficult everything is and making excuses for lack of progress.

Routine

Fifth, the delivery unit leader needs to build routine processes so that the political leader can check that progress is being made. The best approach is regular "stocktakes"—every month, two months, or quarterly—on each key area to review progress and solve problems. The routines driven by the chief minister's stocktakes—informed by good data—are the key to transformation.

Reality

Sixth, and finally, successful delivery units are always informed by reality—through visits to the front line and by being fully in touch with what is happening on the ground. They understand that however good the data is, it doesn't tell you all you need to know.

Conclusion

Is any of this conceptually difficult to grasp? Not at all. At face value, it is no more than common sense. So, the question arises, why don't all governments work like this all the time?

The answer is that governments are complicated—there are always crises getting in the way, departments don't find it easy to collaborate, and politicians can become so absorbed in the presentation policy that the hard yards of delivery are neglected while bureaucrats can find themselves circulating papers explaining why everything is so hard and progress so slow. The whole machine risks becoming comfortable and mediocre. The challenge, therefore, is not conceptual; it's about ambition, focus, clarity, and urgency. Any government can change the way it works for the better as long as it is determined to do so and goes about it in the right way. A good place to start is to set some ambitious goals, gather regular data, and establish routine stocktakes. As with the dramatic improvement in bicycle performance over my lifetime, transforming government is all in the detail.

This leaves one lesson more, the most important of all: never forget the moral purpose of government, which is to enable citizens to live more fulfilled and productive lives. You can ride the most beautiful bike in the world, but it is often admiring the view from the top of a mountain that inspires you to get on the bike in the first place.

—Sir Michael Barber is founder and chairman at Delivery Associates (DeliveryAssociates. com) and author of *How to Run a Government: So that Citizens Benefit and Taxpayers Don't Go Crazy.*

StateStat Dashboards: Measuring Progress

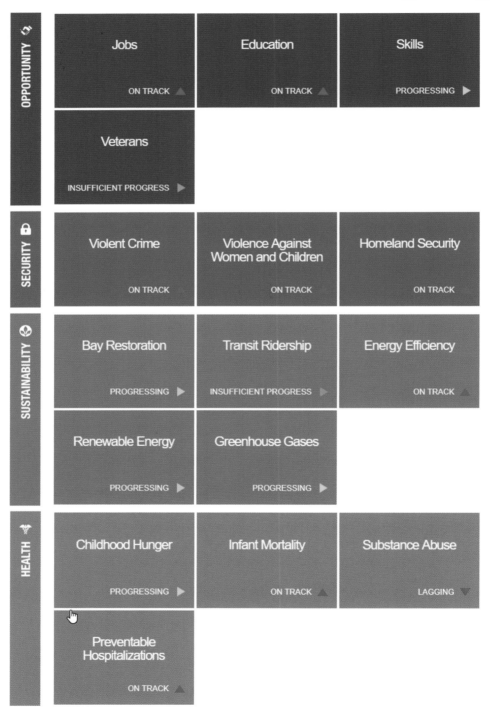

OPPORTUNITY

Jobs — ON TRACK ▲

Education — ON TRACK ▲

Skills — PROGRESSING ▶

Veterans — INSUFFICIENT PROGRESS ▶

SECURITY

Violent Crime — ON TRACK

Violence Against Women and Children — ON TRACK

Homeland Security — ON TRACK

SUSTAINABILITY

Bay Restoration — PROGRESSING ▶

Transit Ridership — INSUFFICIENT PROGRESS ▶

Energy Efficiency — ON TRACK ▲

Renewable Energy — PROGRESSING ▶

Greenhouse Gases — PROGRESSING ▶

HEALTH

Childhood Hunger — PROGRESSING ▶

Infant Mortality — ON TRACK ▲

Substance Abuse — LAGGING ▼

Preventable Hospitalizations — ON TRACK ▲

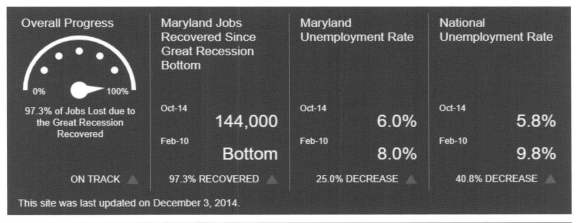

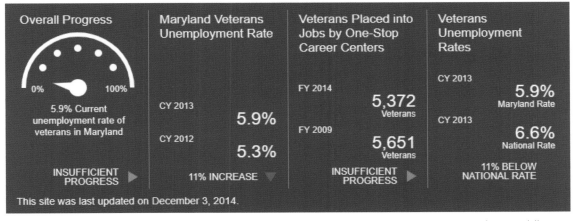

Strategic goals and dashboards became far more important and necessary in the more complex "middle space" of driving progress across Maryland. StateStat worked closely with the governor's office and various state agencies to drive progress toward sixteen strategic goals across the broad areas of opportunity, security, sustainability, and health. A high-level dashboard (left) showed whether we were on track, progressing, or lagging on each goal. Clicking on individual goals revealed more-detailed dashboards for each, such as those for job recovery (top), skills training (center), and full employment for veterans (bottom).

literally—a line on a graph shows the anticipated ramp-up of effort, tactics, strategies, collaborations, and other inputs applied to a problem over time.

When we have a plan and line that shows where we are *supposed* to be going, we can better assess with a second actual line—over the course of time—whether what we are doing is "delivering" the desired result.

Michael Barber was dubbed a knight—presumably for making Her Majesty's government work. And Sir Michael quips that before the innovation of delivery plans with their "trajectories of progress," the most sophisticated tool in national policy implementation was a ruler.

One of the key insights of Barber's approach is to recognize that some public goods or services can only be delivered (or made better) after a series of linked and coordinated actions. In delivering public goods at the local or city level, the links of connected actions—"the chain of delivery"—can be very short, but for goods that we expect our state or nation to deliver, the chain of delivery can be very long and involved.

For example, there are not many steps in the cause and effect of filling a pothole. But there are many, many more links in the chain of delivery of a public good at the state level, such as improving reading scores for third-grade boys and girls. Student-parent-teacher is just the beginning of that long chain. And there are lots of intervening factors like health, hunger, and poverty along the chain of delivery that come into the equation with different degrees of impact for each kid.

If there is a break anywhere in the chain of delivery, the public good will not be realized, and the strategic goal will not be reached. And because the chain of delivery is long—student-parent-teacher-principal-county superintendent-county Board of Education-county government-state superintendent-state board-state government—the chain must be mapped, and the actions expected in each link must be understood, tracked, changed, put back on track, accelerated, and so forth as execution of the delivery plan proceeds and unfolds.

There are also externalities and conditions that must be mitigated against—poverty, health, and nutrition, to name a few in the education mission. And these externalities and their impact on kids won't address themselves. It is important that every link in the human chain of delivery understands their role in reducing these adverse impacts on the learning mission. If kids are showing up to school hungry, whose job is it to make sure school breakfast is provided in the classroom? Who pays? How do we get it done? How do we make sure it happens?

With Michael Barber's help in 2008, we created a Delivery Unit within the StateStat team. Over time, the functions of the unit simply became a part what it took to make StateStat effective at the higher and broader level of state government.

The various coordinating offices of the governor—Children, Youth, and Families; Crime Control and Prevention; Smart Growth; Disabilities; Maryland Energy Administration—connected to a monthly (rather than biweekly) cycle of StateStat meetings. The heads of these legacy "coordinating offices" had a seat at the table—which was actually a big shift. Instead of nipping at the heels of the herd, they sat with equal rank at a collaborative table. They developed closer relationships with the analysts in StateStat assigned to their particular delivery goal or subgoal. They didn't need to beg and plead for the governor to pull large department heads into a special meeting. We were having one every month, and unlike perfunctory cabinet meetings, these meetings were focused on achieving specific goals by a certain date.

We went through a very extensive collaborative process to design delivery plans—with stakeholder buy-in—for every strategic goal. These plans were routinely revised and updated and were posted online for all to see with a link from the Strategic Goals page. Progress (and lack of progress) to a specific strategic goal was the first topic discussed at every departmental StateStat meeting.

As we learned and evolved, and as we searched for better and more effective ways to get things done, the seats around the table grew. Our circles of collaboration expanded even as our focus narrowed in on the strategic goals.

On top of the monthly cadence of departmental StateStat meetings, we added a regular cadence of multi-departmental StateStat meetings. The first was BayStat. BayStat would be followed by meetings focused on veterans, re-entry, public safety, and others. These were short meetings that focused the attention of several different departments on common strategic goals that we could only achieve through deeper and more effective collaborations across the different departments of state government.

And this, too, was a big shift.

Collaboration

When I left the mayor's office, CitiStat had developed into an executive routine. But even with the regular cadence of rotating biweekly meetings, CitiStat was mostly an ongoing series of performance management meetings between the leadership of two bureaucratic hierarchies—the mayor's command staff, and the command staffs of specific city departments.

It was methodical, it was relentless, and it was performance-measured. There was a single source of truth (the data), organized on a common platform (the GIS map), and our operations were made open and transparent to all (a website, the internet, 311, individual customer service numbers, and service

A StateStat meeting in progress, circa 2008. Our cadence changed to a monthly rather than bi-weekly rotation of meetings. Some meetings became far more collaborative by involving multiple departments around the same table. But the method remained the same: timely, accurate information shared by all; rapid deployment of resources; effective tactics and strategies; and relentless follow-up.

guarantees for individual citizen complaints).

All of this was a huge step forward, but we had no routine method for collaborations among and between departments across government.

There were exceptions, such as the eradication of childhood lead poisoning or the cleaning and boarding of vacant houses. DrugStat was another exception that brought health, housing, and employment to the table to make more people successful in recovery. There were also snow and other emergencies that required a level of collaboration among different departments. Some initiatives, like Project 5000—an initiative to take title to thousands of abandoned properties—or the Spring and Fall Cleanups also required a degree of collaborative planning and partnering with foundations,

Federal and State Collaboration

The American Recovery and Reinvestment Act of 2009

In the wake of the great financial meltdown and the recession that followed, our nation elected Barack Obama to his first term as president of the United States.

As the full weight of the recession hit and he was sworn into office, he took a number of actions to stop the free-fall of our economy. One of those actions was the introduction and passage of the American Recovery and Reinvestment Act of 2009, which included an infusion of $1.4 billion through existing federal and state programs to get Maryland's economy moving again.

Critics howled that the almost unprecedented amount of federal spending (to address an almost unprecedented economic downturn) was fiscally irresponsible. And when the bill finally passed, naysayers across the country predicted the huge appropriation of federal dollars would soon result in a tidal wave of waste, fraud, and abuse.

The waste, fraud, and abuse never happened. The reason it didn't happen was because of a new way of governing. The reason it didn't happen was because of new technology, made available to every state in the union. The reason it didn't happen was because of the data, the map, and the method.

And here is the how and why.

Because almost all the federal infusion of cash was coming through existing programs, there wasn't any mystery as to where it would go. The challenge was merely to use GIS and the internet to make its destination clear for more than 300 million citizens who had a stake in seeing it spent well.

neighborhoods, and nonprofit partners outside of government. But these examples were the exceptions; they were not a methodical management routine.

In order to achieve our strategic goals as a state, StateStat increasingly became more collaborative. And it became more collaborative in methodical and routine ways. Multiple departments, in a larger circle, focused on cross-functional ways to better achieve progress toward a particular strategic goal.

The earliest example of this was BayStat (which you'll read about in chapter 10). In truth, the forerunner of BayStat was the Bay Cabinet created in the early 1980s. The Bay Cabinet was a statutory creation inspired by former Governor Harry R. Hughes's passion for saving the Chesapeake Bay. At his urging, the legislature enshrined into law the importance of forcing different departments to meet

To this end, Maryland partnered with Esri to create a template and a website—with open source software available to every state—to track the dollars: from appropriation to spending, right down to the local zip code level.

Maryland was first, but we were not last. Within two short months, every state in the nation was tracking their part of this big federal spending through websites and maps that every citizen could easily navigate. And there it all was—road projects, water projects, school funding, unemployment benefits, and more. Click on any project and every citizen could see a project in their own backyard, a description of the project, the date it was bid, the date of anticipated completion, minority business enterprise/women's business enterprise participation, and more.

There was never any real amount of waste, fraud, and abuse associated with the American Recovery and Reinvestment Act. In fact, there was less marginal waste than there is with most federal programs. Earl DeVaney—the career inspector general charged with overseeing the act—said there was a reason for the lack of waste, fraud, and abuse in such a large federal expenditure: because of GIS technology and the internet, there were 300 million pairs of eyeballs watching where it was supposed to go, and where it actually went.

Effective collaboration—like public trust—requires openness and transparency.

The ability to be open and transparent with the spending of public dollars is not some mysterious search for the Holy Grail. The technology we have today makes it easy. As governor—because of our experience with tracking the deployment of federal dollars from the Recovery and Reinvestment Act—we started tracking the deployment of all state capital budget dollars in the same way so every citizen could see. New companies such as ClearGov and OpenGov are helping states, counties, cities, and towns across America do the same.

It's our money—we should be able to see how it is spent, right? Nothing un-American about that.

together and work together on issues affecting the health of the Bay. What they couldn't put into law was the executive will to convene and drive such meetings.

To this good idea, we added the new technology of the GIS map, the proven methods of a performance management regimen, the structure of delivery plans, and trajectories of progress for the reduction of nitrogen, phosphorus, and sediment. BayStat also used publicly facing dashboards to measure efforts in each of the major riversheds that flowed into the Bay.

Around the table were not only the members of the governor's command staff and the likely departments—Environment, Natural Resources, Planning—but also the secretaries of Agriculture and Transportation. Frequently our head of Intergovernmental Affairs was at the table. Given the importance of county governments in Bay clean-up, their representatives were also present. Very importantly, we had the top scientist from the University of Maryland's Center for Environmental Science there to call balls and strikes on the pollution removal value of individual actions. We also had the dean of the College of Agriculture and Natural Resources at the table, so we could better understand best management practices in agriculture.

As time progressed I reached beyond the circles of government to bring a citizen of wisdom and experience to the table named Russ Brinsfield. He was the former mayor of the Eastern Shore town of Vienna on the banks of the Choptank River. He was a farmer. He loved the Bay. He could speak "farmer" as fluently as he could speak sports fisherman or environmentalist. I invited him to these meetings not only because of his knowledge, but because of his manner and fluency. These were tough conversations. Many of our initial meetings degenerated into turf battles. Chemistry is important for collaboration. Russ's presence at the table and in the dialogue greatly improved our resilience as a group.

One unlikely Bay Cabinet member invited himself to BayStat. And that was Gary Maynard—our secretary of Public Safety and Correctional Services. Gary had taken over one of the most troubled correctional departments in the country. He also had responsibility for the chronically underperforming functions of parole and probation. In his first fifty-two days in the job—with his collaborative

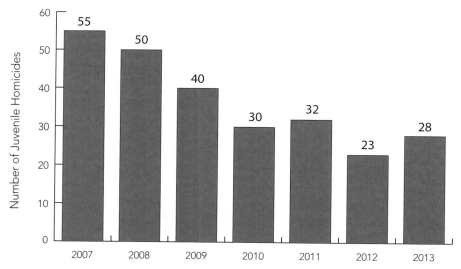

In Maryland, the state runs juvenile services rather than local governments. Performance-measured collaborations and follow-up between social services, schools, courts, police, and juvenile services drove juvenile homicides to record lows.

leadership—we closed Maryland's oldest and most dangerous prison, the House of Correction. The potentially dangerous, complicated, and tightly held operation of moving every inmate out to other facilities in the course of one night went flawlessly. When it comes to collaborative leadership, I have never had the privilege to work with anyone better than Gary Maynard.

One day, Gary asked me if he could attend BayStat meetings. "Of course," I said. "Why?"

"There might be something to the work of Bay restoration that could help me improve the process of re-entry for inmates leaving our prisons and reduce recidivism so they don't come back."

And indeed, there was. We called it "restorative justice." It is a well-known fact that our country has not only the highest rates of violent crime among developed nations but also the highest incarceration rates. We worked in our state to attack that injustice on both ends. If punishment is separation from nature and society, perhaps successful re-entry is about re-connecting back to both nature and society.

Soon, Secretary Maynard was engaging inmates nearing the end of their sentence in important work to restore the health of the Chesapeake Bay—building cages for our oyster replenishment project, working at the facility that produces the spat for oyster reefs, building artificial reefs, and restoring stream buffers along public lands by planting a million trees, as well as innumerable acres of grasses across Maryland. When the supply of saplings became short, Gary even started his own nursery on the state prison grounds, enabling us to grow our own native trees. He also created ways for inmates to glean farmers' fields of produce that stocked the shelves and freezers of the Maryland food bank.

All of this collaborative work contributed to the strategic goal of restoring the health of the Chesapeake Bay. But it also helped reduce recidivism by 20 percent, improving the likelihood of a successful re-entry. The restorative justice work was not everything we did to improve re-entry and reduce recidivism, but it was an important part of it.

An aerial view of Annapolis and the Chesapeake Bay in Maryland.

Place Matters

We did not set a strategic goal that to some might have appeared a likely singular fit for our Department of Human Services—that department responsible for administering welfare, food stamps, welfare to work, and protecting vulnerable kids. But to our secretary, Ted Dallas, and his predecessor, Brenda Donald, all the goals did fit—jobs, security, opportunity, health, the environment.

One strategic goal that clearly could be impacted was our goal of reducing the child victims of violent crime by 25 percent within four years, including homicide victims.

Often, when confronting a whole range of chronic problems, we hear people say from experience, "If we only spent a couple more dollars up front, we wouldn't be spending all this money in the back end now that the problem has become worse." In this case, we knew from fatality reviews of juvenile homicide victims that the vast majority had spent some time in our foster care system before being murdered.

But system change—even change that eventually saves money and lives—requires money up front, and relentless follow-up over time to make the change happen. Attracting a serious partner to finance and support that change requires a serious performance management regimen inside government that will guarantee the collaboration succeeds— that we actually do what we set out to do. After all, who wants to dance with someone who doesn't even want to get out of their chair?

By governing well, we were able to attract that serious partner from outside of government.

"Place Matters" was a partnership between the State of Maryland and The Annie E. Casey

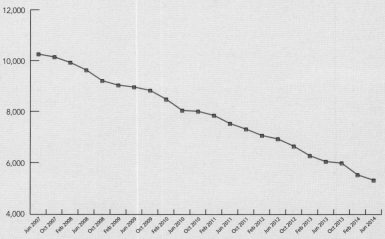

The number of children in out-of-home placement—foster care or group homes—was driven down by 48 percent from 2007 to 2014. It was one of the largest decreases of any state in the nation. A host of actions that focused on healing the constellation of problems in a child's family home made all the difference. This reduction accompanied—and, I believe, contributed to—a 50 percent reduction in juvenile homicides over the same period that we learned how to better heal families.

Foundation to heal the family homes of children reported at-risk instead of simply removing the child from the home and placing them in foster care. It was a model of leadership, partnership, innovation, collaboration, and delivery. It contributed greatly to the major strategic goal of reducing juvenile homicides and harm. And it worked.

One of the greatest stories of reforming a troubled big department is the story of the turnaround at the Maryland Department of Human Resources—the department responsible for Social Services, but also the protection of our most vulnerable children who are being raised in some of our most fragile family settings.

The five interrelated tactical goals of the reform effort were:

• Keep children in families first and reduce the numbers of children living in congregate care and group homes.
• Maintain children living in their communities.
• Reduce reliance on out-of-home care.
• Minimize length of stay in the foster care system.
• Manage with data and redirect resources to family healing and prevention.

These interrelated strategies, or leading actions, involved improving the rapid deployment of services to a vulnerable child's family, improving our effectiveness at family reunification, adoption, and guardianship. Each of these leading actions had a host of smaller changes with these efforts but the compelling scoreboard was clear.

In any given week, there were more families being united or more being broken up. Either there were more kids being placed in foster care or more kids being placed in guardianships and adoptive homes.

By focusing first on healing a vulnerable child's home and family situation rather than simply ripping the child out of his or her home and sticking them in a group home or foster care, we reduced out-of-home placements by 50 percent in eight years. This also happens to be close to the 45 percent reduction in child deaths in Maryland. More than 19,300 children were placed in permanent rather than foster homes because of increased efforts at family reunification, guardianship, and adoption.

We also exceeded our goal in reducing child victims of violent crime by 34 percent in our first four years. And by the end of eight years, we had cut juvenile homicides by 50 percent.

Place matters. And so does every life.

What I learned about performance management at the state level is that performance management without cross-functional collaborations can only get you so far. Stated more positively, if you want to achieve strategic goals at the state level, structured collaboration is every bit as important as the routine of performance management.

We never stopped making changes and adaptations to how we functioned as an administration. The goals rarely changed. But the ways to achieve them were always changing. And it was always very much a work in progress.

That brings me to innovation.

Innovation

Innovation should be everybody's job. And yet, if you have ever tried to pitch high-level people in

Starting Up Your Own Stat System

A few years ago, the US Conference of Mayors asked me for a ten-step process that new mayors might follow in setting up their own CitiStat system. Years later, the National Governors Association asked me to prepare tips for incoming governors, both Republicans and Democrats, to use to start their own performance management or "stat" system. With minor tweaks, the steps for beginning are really the same.

Many well-intentioned executives fail to implement an effective performance management system because their senior staff members kill it before it begins. The staff think that they are saving the executive from the political pitfalls of public goal-setting, public performance measurement, public openness, and public transparency.

Don't let your (especially) experienced staff kill implementation with excuses, such as "We can't afford to do that right now."

Or, "We tried that, and it didn't work."

Or, "We are already doing that."

Or, "We can't afford to waste time in meetings."

Or, "To do this, everyone would have to stop working on other priorities, and it will take months."

Only one person was elected chief executive. Tell your staff to do it or step aside so you can find someone who can. You weren't elected to manage people's feelings; you were elected to achieve results.

government about new approaches, new technologies, new ways to get things done, you can practically see the drawbridge come up.

It's not always a conscious contempt for new ideas, a disdain for vendors, a feeling that citizens don't understand, or a dread of immoveable bureaucracy, although it can be all those things. Most often, it is simply the crush of the daily work load and the weight of the latest crisis. Add to those pressures another round of budget cuts, and it is easier to see how even the most open-minded managers can conclude they are too busy doing their job to listen to a bunch of "crazy ideas" about how they might do that job better.

In 2011, we created a new position of chief innovation officer in the second floor of the governor's office. It was a first for us as a state. Many states and cities have now followed suit.

Why a chief innovation officer, and what does he or she do?

Ten Steps to Starting Your Own System
1. Prioritize the order of agencies to be rolled into the stat process. The first meeting should happen six to eight weeks later, and a new agency should be added every two to four weeks in the rollout plan.
2. Announce the date of the first meeting to give everyone a hard deadline against which they must work.
3. Preselect a few easily understood/easily communicated wins. Something such as eliminating a backlog (such as DNA analysis, food stamp applications) or a service improvement (pothole guarantee, graffiti removal, and so on).
4. Designate a director and one analyst. They can build from there, and in the meantime, rely on an analyst from the budget office and a GIS person from their central IT or a GIS-intense agency (such as Transportation or Planning).
5. Designate a physical space for the meetings. At a minimum, seating for fifty, two large screens, and a podium. Spend the money to make it permanent—wall board, projection room, installed furniture, and so on—not just folding chairs, pop-up screens on tripods, folding tables, and other temporary solutions.
6. Complete an overview of the existing performance measurement infrastructure. At a minimum, look at the agency monthly budget reports, annual budget testimony, work order reports, audits, and so on. Look for usable baselines. Tell incumbent managers, "give us everything you use to manage your agency."
7. Take the results of this initial data review and juxtapose the data points against what's available from an established stat system with an eye toward looking at gaps. If you don't track some things, but they do, probe those gaps to find deficiencies.
8. Meet with the existing or incoming leadership teams of the agencies prioritized for the stat rollout to get their feedback on measures/areas of focus.
9. Take a field trip to a city or state with an established stat system to see meetings and observe. Remember not to get discouraged; it takes a few months to get into a rhythm, manage the data submissions, and so on.
10. Spend a little time up front baselining financial and survey data so you can capture your improvements later.

Between the time, angst, and tears we were spending on cutting the operating budget every month as the ongoing recession drained state revenues, not to mention the energy we were expending on continuous improvement, it seemed to me that no one on the second floor (or in the cabinet secretary's offices) had the time and energy to search for, listen for, and be open to more innovative ways to get things done.

In our final three years in office, the US Chamber of Commerce would name Maryland the number one state in America for innovation and entrepreneurship. Part of that was the result of actions we took to improve education. But a part of it was the recognition of the innovation assets that call Maryland home: the National Institutes of Health (NIH), US Cyber Command, National Institute of Standards and Technology (NIST), NASA Goddard Space Flight Center, Johns Hopkins University—the list goes on.

Tapping into that font for the sake of achieving our strategic goals is what the chief innovation officer was tasked to do. Vendors, stakeholders, front-line managers, and workers are always tossing ideas up. Whose job is it to catch them, or to do the due diligence required to determine if they might have some value for the mission? Whose job is it to cast about for private-sector innovations to help us solve problems that we encounter?

In an ideal world, those things should be everyone's job. At the enterprise level—to advance the attainment of our strategic goals—it became the job of the governor's chief innovation officer to perform both an ombudsman role and a lookout role; to attend StateStat meetings, to look over the horizon, to listen to what hard-working people were trying to accomplish, and to keep ears open for news ways of getting it done.

Making Changes While Still Following the Most Important Rules

There were some significant differences in bringing this new way of governing to a state compared to bringing it to a city. Those differences had to do with scale, size, and breadth of mission. They had to do with the mid-level nature of state government and a lack of visibility when it comes to the challenge of delivering public goods on common problems—problems that can only be addressed if multiple authorities of government work together.

Collaborative leadership is sometimes difficult for leaders who are steeped in the experience of staying in their own lane. Right after I was sworn in as governor, I asked our Maryland State Police to provide for me every morning—as the Baltimore Police Department had for me as mayor—a one-page sheet showing homicide and violent crime numbers statewide compared to the same date last year. What I received were the numbers for the mostly rural areas of the state where our state police had primary responsibility for first response. Hmmmm. I told them I wanted the whole state. Fulfilling that request took four months and a few tries. It wasn't that the numbers didn't exist. They existed in lots of different places. No one in a position of authority had ever asked to see them all in one place.

Our state police chief could understandably resist the notion that his department—steeped in eighty years of tradition—was somehow responsible for local crime numbers. But when we realize how much more effectively our state police could partner with local agencies on intelligence sharing, warrant service, crime lab effectiveness, advances in DNA technology, digital fingerprints, networked license plate reading technology, face recognition technology, and other areas, the pathways of collaboration become clearer.

We had to adapt. We had to change to be effective. Over the course of time and experience, a few key differences clearly emerged to distinguish StateStat from CitiStat. The most important of these attributes were the result of conscious and intentional changes. Among them—the necessity of setting strategic goals, the concept and disciplines of delivery, the importance of cross-functional collaborations, and the imperative of greater openness to innovation.

But the most important rules still held.

We started and we never stopped.

We lifted up the leaders.

And we led.

Learn & Explore

Data-Driven Delivery: Lessons from the O'Malley Administration of Maryland
This report is from the Institute for Government on using StateStat to measure performance and deliver results.

Road Ready Operations Dashboard

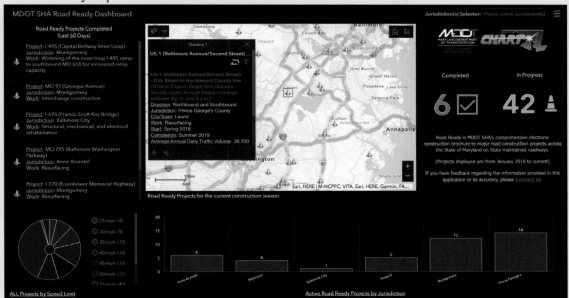

This dashboard from the Maryland Department of Transportation shows State Highway Administration projects completed in the last sixty days. When capital budgets and projects are mapped, measured, and tracked in a way that everyone can see, it's amazing how many more get done on time and on budget.

For links to these and other examples, exercises, and resources, visit SmarterGovernment.com and click chapter 7.

8

Improving Education

Improving public education is the key to making our children winners in this fast-changing economy. The more our people learn, the more our people earn and the more inclusively our economy grows.

We know what works—the challenge is to do it.

It's all about pedagogy.

Learning More to Increase Opportunity for All

When I was elected mayor of the City of Baltimore in 1999, not one of the grades in our public elementary schools was testing majority proficient in either reading or math. We were the poorest jurisdiction in the state, and we had, by far, the poorest-performing school system. By the time I left the mayor's office in 2007, grades one through six were scoring majority proficient on state reading tests, and grades one through five were scoring majority proficient in math.

In 2009, when I was serving as governor of Maryland, our state for the first time ever—and in the middle of a recession—was named the number one public school system in America by *Education Week* magazine. It was a distinction we held for five consecutive years.

These accomplishments did not happen by accident. They were the product of hard work, persistence, and leadership at many different levels of a complex system—from the statehouse to the schoolhouse. And quite honestly, we did not set out to win any particular awards. Our goal as a state and as an administration was to improve student achievement and college and career readiness by 25 percent by 2015.

The difference between a dream and a goal is a deadline.

We never made perfect, but we did make better. And every state in America can also do better.

A truly inclusive economy—one with better opportunities for all of our kids—requires better skills and better education in every generation. It cannot be otherwise.

There are exceptions to every rule, but generally speaking, the more a person learns, the more a person will earn. This rule holds true for individuals, and it also holds true for generations. It is true for families,

In Baltimore, our Believe campaign (see chapter 3) transcended the challenges of our crime and drug problem. It engaged citizens at every level, including our youngest ones, in some of the most important work of all—achieving a quality education. Mayoral message to every class—"I love you, I am proud of you, and I need you." Presence, commitment, and clarity of communication.

and it is true for countries. Many developing nations have now put this American economic truth into practice to grow the size and strength of their own middle classes.

Our parents and grandparents understood this truth well. They also understood another important truth: our economy is not money; it is people—all our people.

If "opportunity for all" is the promise at the heart of the American Dream, a good education is the personal key to unlocking that promise.

A Time for Progress in Public Education

In more recent times, as the pace of technological change has accelerated, the pace of educational progress in our own country has not. Public institutions in a democracy can be slow to change. They can also be resistant to change—especially when authority is spread over multiple levels of government and multiple layers of public responsibility. What some citizens see as political accountability, some school board members and college presidents see as political interference.

We can debate, scapegoat, and blame. Or we can make progress. The choice is ours.

We know what works. We know what we need to do more of. And we know there are proven ways to close achievement gaps between kids from poor families and kids from wealthier families. Many of them involve improving the learning environment and well-being of children.

We know this works, because some places are doing it.

In this chapter, we will focus on the importance of a few key frameworks when it comes to improving public education for children throughout an entire state. Those frameworks are collaborative goal setting, delivery, leading actions, and feedback loops.

The Necessity of Collaborative Leadership

Improving public education throughout a state is not a paint-by-numbers exercise. It requires a degree of collaboration and persistence that eludes many leaders and many people. It is not impossible; in fact, it is wholly possible. And we have never had better technology and tools available for mapping, measuring, and tracking progress right down to the level of the individual student than we do right now.

But it does require a lot of collaborative work. And just because a governor cares doesn't mean the local school boards or college presidents listen or care to follow anyone else's orders or urgings.

When I was elected governor of Maryland in 2006, our school system wasn't bad; we were better than most states. But better than most wasn't good enough—especially if your family lived in one of the underperforming local school districts of the state.

Our state constitution gives the governor considerable authority, but it does not grant the governor authority to dictate education policy. Nor does our constitution give the governor the authority to order the sort of follow-through at the district level that turns policy into higher achievement for students. Some of the major stakeholders have terms of office that are actually designed to outlast even a two-term governor.

Education progress therefore requires collaborative leadership, and collaborative work. It requires: vision, belief, presence, the convening of stakeholders, and relentless focus. Creating the space to ask more questions about what works, what does not, and what might work better. Listening. Being open, honest, and respectful of the perspectives and responsibilities of others. Being accountable to the group effort by delivering your piece of the enterprise with integrity.

Easy to say. Harder to do.

So how did we do it? In 2007, we had a fair amount of data on student achievement—state

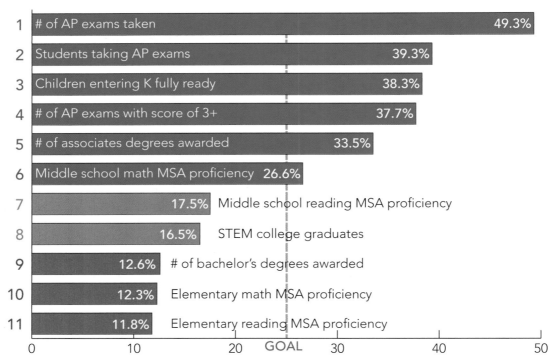

1	# of AP exams taken	49.3%
2	Students taking AP exams	39.3%
3	Children entering K fully ready	38.3%
4	# of AP exams with score of 3+	37.7%
5	# of associates degrees awarded	33.5%
6	Middle school math MSA proficiency	26.6%
7	17.5% Middle school reading MSA proficiency	
8	16.5% STEM college graduates	
9	12.6% # of bachelor's degrees awarded	
10	12.3% Elementary math MSA proficiency	
11	11.8% Elementary reading MSA proficiency	

Our statewide strategic goal for education was to improve college and career readiness by 25 percent by 2015. The component parts of that aggregate 25 percent were determined and measured by our success of the eleven education goals displayed above. These were lagging indicators, which told us whether or not an array of new tactics and strategies were working. Green = the goal has been met or exceeded; orange = we are making progress, but more work is needed; and red = insufficient progress.

standardized test scores, graduation rates, advanced placement exams, enrollment, and degree attainment in community colleges and our so-called four-year universities where it took the average student six years to graduate.

We collected all this data and more, and we translated it into maps and graphs. We put these realities on one map for all to see. We set a big strategic goal—to improve student achievement and college and career readiness by 25 percent within four years—and we agreed on a series of subgoals across the continuum of the learning process from pre-K through college. We created a few basic dashboards and scorecards to measure and compare progress across time and across jurisdictions and institutions. We started bringing responsible stakeholders together regularly to focus on student achievement, and we never stopped.

The strategies that follow are laid out sequentially only because that is how a text flows. In actual practice, we took these actions simultaneously, not sequentially. We were building even as we were sailing. Changing tires on a rolling car. Planning for contingencies that none of us could guarantee. Managing for results even through a deep and unpredictable recession.

The dialogue, the search for better ways, and the pursuit of more effective actions to achieve our goal never stopped. Even when the 2012 deadline rolled around, we knew we would follow it by setting a higher goal with another deadline. And we did.

The Framework for Education Progress

First—with the convening power of the governor—we formed the P-20 Leadership Council. This did not exist in law or in the constitution. We did it by executive order.

The P-20 Council was a collaborative circle—appointed by the governor—of responsible leaders and stakeholders from every stage of our public education process—from pre-K to elementary school, from middle school to high school, and from high school to community college to four-year universities. Hence, the name: P-20.

Public, private, non-profit, business, and union leaders were all brought together, meeting regularly to set goals, measure progress, and take action on the drivers for improving student achievement. This idea was borrowed from former Governor Mark Warner in Virginia.

Second, after a lot of collaborative discussions, I set the state goal of improving student achievement and college and career readiness by 25 percent by 2015. This became a key strategic goal for our entire state—not just for our state Department of Education. And with the P-20 Council, we developed a plan with objective criteria and measures to tell us whether we were achieving our goal.

Third, we invested more—37 percent more—to improve public education in every part of our state. Each of us might take our kids to a "county" or "city" school, but 40 to 60 percent of that local school district's budget is paid for by our state tax dollars. Maryland was one of only about a dozen states that also invested hundreds of millions of dollars every year into school construction and renovations. By the end of my administration, forty-nine cents of every state tax dollar was going to fund education. This increased level of investment allowed us to expand full-day kindergarten statewide and to greatly improve early childhood education. Both of these actions had a big impact, especially in our poorest school districts.

Unlike many other states, we treated our teachers with professional dignity and respect. Borrowing from Governor Michael Easley in North Carolina, we conducted a regular survey of all our teachers on a biannual basis to ask questions such as:

- What's working in your classroom?
- What professional development would be most useful in your classroom?
- What can we do to improve the learning environment in your classroom?

The dignity of every child's potential, respect for the teaching profession, mutual accountability, ownership—these values weren't necessarily in the text of every survey question, but they were certainly communicated between the lines.

And our teachers responded to this sort of collaborative leadership with professionalism, partnership, and a renewed commitment to excellence—often finding ways to improve their own teaching skills in science, technology, engineering, and math (STEM) before a more formal upskilling regimen could be crafted at a county or state level.

Next, we embraced the new and improved Common Core Curriculum so that our children would have the skills they need to compete and win in a global economy. This was not done without controversy and a lot of listening. There were implementation issues and, together, we had to work through them.

Together, we also increased state investments to make college more affordable for more families, rather than more expensive. Maryland was the only state during the recession to go four years in a row without a single increase in college tuition. This was not by chance; it was a choice. It was a leading action that was consistent with our larger goal of increasing student achievement and college and career readiness.

Like the pursuit of other big important goals, there were certain essential components and practices for success. Among them: common platforms, alignment of effort and standards, a compelling scorecard that all could see, and relentless follow-up. We developed a statewide data system to track the progress of every student throughout their educational journey from pre-K to post-secondary education in college and beyond. This longitudinal data system had unique identifiers for every student so their progress and needs could be tracked, regardless of how many times they transferred schools or moved from one county school district to another.

Finally, we built the trust required to earn the support of some really outstanding leaders among the county and city superintendents, as well as some really outstanding principals and teachers throughout our state.

We never made perfect, but with openness, transparency, collaborative leadership, and the setting of goals with deadlines, we did make nation-leading progress.

Collaborative Goal Setting and Transitional Readiness

Together we settled on one over-arching goal for public education in our state—to improve student achievement and college and career readiness by 25 percent by 2015.

Simply measuring the endpoint of the learning process would not cut it. Stakeholders recognized that their ability to deliver improved results at their particular stage of a student's educational career depends on the effectiveness of the learning mission in the years before elementary school, middle school, high school, or college.

Understanding the importance of "transitional readiness," we looked to establish key performance metrics along the longitude or continuum of learning—from pre-K to college or career. After a lot of collaborative dialogue and conversation, we settled on the following indicators—or lagging measures of progress—across the pre-K to college or career arc of learning:

- The percentage of children entering kindergarten who are ready to learn
- Elementary school reading and math proficiency
- Middle school reading and math proficiency
- High school students taking Advanced Placement (AP) exams
- Number of high school AP exams taken
- Number of students who score 3 or higher on AP exams
- Number or associate degrees awarded
- Number of bachelor degrees awarded
- Number of STEM degrees awarded

Our goal—and therefore, our intention—was to drive progress on each of these lagging, and mostly annual, indicators by 25 percent within four years. The effort of every student, parent, teacher, school, and district was important to the achievement of the statewide goal.

Collectively, these lagging measures of student achievement and readiness were easily displayed as a dashboard by which we could measure our success or failure. Nobody wants to be in the red; everybody wants to get to green. Outcomes and pace of progress would vary from one measure to the next and from one school district to another. The map would allow us to see which school districts were making sufficient progress toward our goal, and which ones were not.

We also understood that we could not move the numbers simply by getting together to look at the dashboard once a year. Each one of these outcomes required a focused strategy with its own set of leading actions. And for the most part, the follow-through and execution of tactics and strategies—and improved pedagogy (the theory and practice of teaching)—would depend largely upon independent local school boards and universities.

We needed to create a faster cadence of accountability than annual assessments and graduation rates. So we did. We needed to focus on leading actions year-round rather than lagging indicators at the year's end. So we did. And all this required a different way of getting things done.

Governing for Results: Improving Education

What did we accomplish in Maryland?

- From 2009 to 2013, Maryland's public school system was ranked first in the nation for five years in a row by *Education Week* magazine.
- By 2013, student achievement and school, college, and career readiness was improved by 31 percent over 2006, surpassing our 25 percent goal.
- As of the 2013-2014 school year, 82 percent of Maryland's students entered kindergarten "fully ready," a nearly 38 percent increase over 2005-2006.
- During the 2012-2013 school year, a record 85 percent of Maryland seniors graduated from high school. During the same period, the dropout rate was reduced to 9 percent—a 25 percent reduction from the 2009-2010 baseline dropout rate of 12 percent.
- The number of AP exams taken by students increased from 65,700 during the 2005-2006 school year to 108,471 during the 2012-2013 school year—a 65 percent increase—and the number of students achieving AP test scores of 3 or more increased by 56 percent.
- During the 2012-2013 school year, Maryland awarded 46,080 associate's and bachelor's degrees, a 29 percent improvement from 35,625 degrees awarded in the 2006-2007 school year.
- Enrollment in STEM-related career and technical education (CTE) programs increased from 1,195 during the 2005-2006 school year to 17,793 during the 2012-2013 school year, representing almost 7 percent of total high school graduates.
- In 2012-2013, more than 77 percent of CTE graduates achieved an industry-recognized certificate or licensure compared with just 35 percent during the 2008-2009 school year.
- In 2013, the number of STEM college graduates increased by 37.1 percent over the 2005-2006 school year.
- By 2015, the American Community Survey reported that 46 percent of Marylanders between the ages of twenty-five and sixty-five had earned a bachelor's degree or higher. This was an increase from just 36.9 percent of Marylanders in 2005.

Delivery Plans and Processes

Somewhere along the course of my own transition from mayor to governor, I came across the work, experience, and writing of Michael Barber in the United Kingdom. As mentioned in prior chapters, Michael, or Sir Michael now, had led the domestic policy efforts of the government of Prime Minister Tony Blair in Britain. One of Michael Barber's lasting and most widely adopted contributions to modern public administration is the concept of *delivery*.

Like its American cousin, CompStat, delivery in a nutshell is the practice of setting goals with deadlines, of measuring and tracking performance, of repeatable routines and short, regular "stocktake" meetings to assess progress and make adjustments to tactics, strategies, and plans. It requires a dedicated team acting with the authority of the chief executive and the constant following and adjusting of clear plans for delivery of any public good.

While CompStat originated at a municipal level within a paramilitary organization having clear levels of rank and authority (the New York City Police Department), delivery was developed at a national level where the degree of cross-jurisdictional complexity and cross-departmental collaboration required for delivering a public good creates a different level of managerial challenge. At this level, the need for the concept of a delivery plan—with clear trajectories and clear responsibilities—becomes essential if one is to have any hope of delivering progress in a politically meaningful time frame.

I once heard Michael Barber describe the difference in this way:

> "When it comes to things we expect of our government, there is a chain of delivery required for providing any public service or public good. At the city or local level, the chain of delivery doesn't really require a plan as much as it requires action.

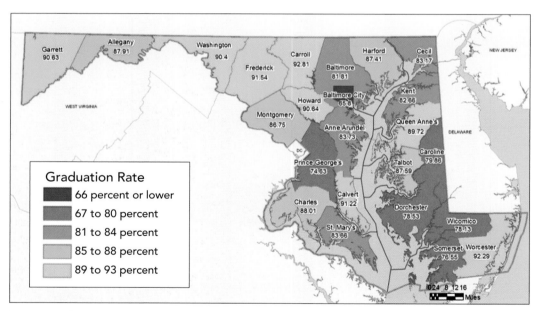

A SchoolStat map of graduation rates across Maryland, by county, for the 2010-2011 school year. We could go to the individual high school district and frequently did. With the development of the longitudinal data tracker, we could also go to the individual student level. There is a difference between No Child Left Behind, and enveloping the system and practice that makes sure every child succeeds.

"A citizen might call in a complaint about a pothole. That is the first link in the chain. The city takes the complaint and dispatches a crew. That's the second link in the chain. The pothole crew from the Transportation Department finds the pothole and fills. That's link three. One, two, three, the service is delivered.

"But when it comes to improving third-grade reading levels of boys and girls throughout an entire country or an entire state in the United States, there are many more links in the chain. And they all need to be mapped and understood by several layers of responsible leaders in that chain of delivery if we are to have any hope of delivering better results.

"First there is the student—the third-grade girl or boy. Then there are their parents, or parent, or grandmother, or foster parent(s). Then the teacher. Then the instructional leader of the school—who might or might not be the principal depending on the size of the school. Then there is the district or county instructional leader, who might or might not be the superintendent; again, depending on size of the size of the school district, the instructional leader could be the assistant superintendent. Then there is the local school board, and the local county government who usually has appointment, funding, and oversight responsibilities. Then there is the state school board and the state superintendent. Then the state legislature with its committees of oversight on education, appropriations, and budget. By the time you get to the governor, there are twenty or more links in that chain.

"And that's just for instructional improvement. Add on top of that the actions necessary to improve the health and well-being of children living in poverty, and the learning environment of classrooms in school districts that have a hard time affording books and paying for preventive maintenance of really old buildings. Now you have a braided chain of delivery."

There aren't a million links—not even a thousand, or a hundred. But there is a level of complexity and collaboration that makes having a delivery plan a must.

A good delivery plan makes plain which leader or leaders are responsible for every link of action in the chain. A good delivery plan distinguishes the lagging indicators from the leading actions. A good delivery plan lays out the strategies. A good delivery plan maps and makes visible the chain of actions required for delivering a public good. A good delivery plan charts the trajectory over the space-time continuum,

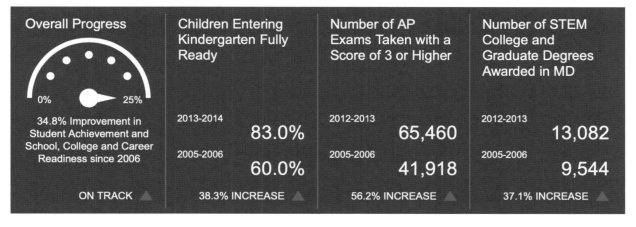

A StateStat dashboard showing progress towards improving college and career readiness by 25 percent by the end of 2015 (as of July 1, 2014).

so progress on any individual strategy—and the actions that make up the strategy—can be regularly assessed for their effectiveness and on-time delivery.

The elements of a good delivery *process* include a dedicated team backed by the authority of the chief executive. A good delivery *process* focuses the stakeholder group on trajectories of progress— trajectories that set expectations for progress over the space-time continuum. And a good delivery process convenes regular stocktakes of responsible stakeholders to assess progress, adjust actions, and ask the essential questions about what works.

Following the delivery process is the key. Constant updating of the delivery plan is a must. No plan ever survives the first engagement.

Like CompStat, CitiStat, and StateStat, you start and you don't stop.

Leading Actions: State and Local

There is a big difference between lagging measures like standardized test scores and leading actions like better training for new teachers.

Lagging measures are nouns. Leading actions are verbs.

Lagging measures tend to be easier to score, while leading actions involve some assessment of the effectiveness of technique or impact of scale.

Some of the leading actions we took to improve education were statewide initiatives that impacted every jurisdiction. Among them: guaranteeing adequate funding for every district regardless of income through increased state funding, mandating and funding full-day kindergarten, reforming disciplinary policies to reduce out-of-school suspensions, mandating an environmental literacy requirement as part of the state curriculum, creating a longitudinal database for tracking individual student achievement, holding down the cost of college tuition, and implementing Maryland's STEM Innovation Network (STEMnet)—an online platform that provided tools, resources, and workforce partnerships to strengthen STEM teaching and learning within the state.

Other actions required decidedly local execution and local implementation. Among them: better teacher training, pay, recruitment, and mentoring of new teachers; better recruitment and training of school principals; making sure children who qualified for free or reduced breakfast actually got fed in the classroom before the school day started; improving time and access to training for teachers who wanted to improve their skills for teaching STEM subjects; and investing in the creation of quality career and technical education (CTE) pathways in more of our high schools.

Beyond the state investment—which was substantial—our greatest tools for driving progress were to lift up the leaders, show citizens and parents which school districts were progressing and which ones were not, bring responsible leaders and professionals together regularly to praise those who were making progress, and, in turn, lift up those leaders in the eyes of their colleagues.

We looked for every opportunity to do just that. To lift up the leading students, teachers, principals, and county superintendents who were making the greatest progress. To visit schools and engage in conversations about their pursuit of the important state goals at their school. To praise, to share best practices, to identify best practitioners. And when necessary, to push and to prod.

Feedback Loops

In any complex system, there are feedback loops. For a broad endeavor like improving public education for an entire state, almost all our lagging measures were annual measures. In fact, the statewide teacher survey was only conducted every *other* year.

The most valuable feedback was not the feedback of the annual lagging measures of kindergarten readiness, test scores, and degree attainment. It was the conversations that took place at the local district level, at the school level, and at the P-20 meetings, monthly SchoolStat meetings, and quarterly "stocktake" meetings as we focused on the actions necessary to drive the lagging measures in the right direction.

In our state, each of the twenty-three counties and the City of Baltimore were their own singular school district. Therefore, the good news was there were only twenty-four school districts in total. But each superintendent had their own management style. Some were more effective than others. As our time together progressed, each of them saw what we were watching and measuring from a statewide perspective. Each of them knew that our over-arching goal was to improve college and career readiness 25 percent by 2015, and the subgoals that defined success. And each of them understood they would be judged professionally by stakeholders in their own counties on whether they were making progress toward our goal.

The local superintendent who I believe did the best job of structuring an ongoing, evidence-based dialogue around the practice of teaching and learning was Jerry Weast of Montgomery County.

Montgomery County SchoolStat: "It's All about Pedagogy!"

Montgomery County, Maryland, is the largest county in our state, and by many measures, it is also one of the most diverse—economically and racially. It also boasts the largest number of learners of English as a second language of any elementary school district in our state. And the first languages of these young students are drawn from the diversity of the globe.

Fortunately, Montgomery County also has one of the higher median incomes of any county in America, and a tax base that is healthy and growing. But schools superintendent Jerry Weast saw

Governing for Results: M-Stat

Under district Superintendent Jerry Weast, Montgomery County Public Schools (MCPS) adopted the "Seven Keys to College Readiness" to ensure that 80 percent of MCPS students would be prepared for college and career ready by 2014. To monitor and measure progress toward this ambitious goal, MCPS adopted M-Stat, based on the NYPD's CompStat process. The results were dramatic, as shown between 2001 and 2009:

- The proportion of students who attained a grade of C or higher in Algebra 1 or another high-level mathematics course rose from 43 to 66 percent, an increase of 23 percent. The increase was even higher for Hispanic students (a 30 percent increase) and African American students (a 26 percent increase).
- AP course participation more than tripled. AP course performance (judged by earning a 3 or higher on at least one AP exam) doubled, and the number of African American graduates who earned such a score in 2009 was more than three times the state average and five times the national average.

clearly that something else was growing, and it wasn't good—there was a growing achievement gap in educational outcomes between wealthier kids and poorer kids, between white kids and black kids and Latino kids.

He threw all his considerable leadership and political skills into attacking the problem head-on. He surfaced the growing problem in town hall meetings, where he displayed a large GIS map of the entire county broken down into just two zones. The Green Zone was where students were achieving at above-average levels, and the Red Zone was where students were performing below average.

Then, he started doing something that few local superintendents have had the guts to do. He not only shared all school performance data openly with the public, but also started bringing principals and instructional leaders together in a cadence of accountability—short meetings, on a regular rotating basis—so everyone involved in the mission of learning could figure out what was working and what was not.

One afternoon in 2005, I found myself sitting at the back of a specially renovated room that looked very much like a rounder version of the our CompStat or CitiStat rooms in Baltimore. Only, in this room at Montgomery County school headquarters, I was a witnessing a real, live, full-on SchoolStat meeting at the county level. Across our country, this is where the vast majority of the operational decisions and instructional work actually gets done.

They called their new process, M-Stat, short for Montgomery County Schools Stat. And yes, schools superintendent Jerry Weast had modeled his system on the NYPD's CompStat, just as we had done for CitiStat in Baltimore. Weast imbued the M-Stat process with the Baldrige problem-solving method (plan/do/study/act) and away they went—up the path of continuous improvement.

Superintendent Weast started, and he never stopped. He lifted up the leaders. He paired different leadership teams from different schools to learn from one another.

At one M-Stat meeting, for example, he might have leaders from one school share their success in reducing high school drop-out rates with the leaders of another school that was struggling with the same problem. This CompStat for education was complete with maps, charts, and graphs. County school leadership gathered around one side of the circle, and school leadership teams gathered around the other side of the circle. And, together, they focused on the latest emerging truth. They asked the questions about what works.

I was both mesmerized and very annoyed. This was exactly the sort of management regimen that, as mayor, I had been prodding, begging, and pleading with my own independent city school system to implement. They got as far as building maintenance and finance but never wholly embraced it. The best they could ever muster were a couple of quarterly "*faux*-stat" meetings.

But here it was—live, happening, and well-led. Repeatable routines, agendas, notes being taken on commitments made. Timely, accurate information shared by all. A constant, open search for the best tactics and strategies in teaching.

At the meeting, I quietly observed, closing the achievement gap was the driving theme. Montgomery County had come up with a new diagnostic tool based on past experience that gave the county a better idea of the minority kids who could succeed at the AP level but for a number of reasons—some related to process, some related to culture, some related to unintentional gatekeeping and screening—were not actually getting into AP classes. They called it the Honors/AP Potential Identification Tool (HAPIT). They provided training throughout the school system for principals, teachers, and counselors. It appeared to work. More minority kids were taking and passing AP-level courses. More schools started doing it.

"There was powerful evidence that gap-closing strategies not only existed but they worked in the

Red Zone (lower-performing) schools," said Donna Hollingshead, one of Weast's staff who worked to drive the M-Stat process. "The data showed some Red Zone schools outperforming the Green Zone schools. These practices were documented and distributed to all schools. This documentation provided principals with the proof they needed to take back to their own schools and move their sometimes intractable staff to reorganize and expand their repertoires to increase student achievement—closing the achievement gap while still hitting or closely approaching district targets."

Jerry leaned over and whispered to me in the middle of this particular meeting: "It's all about pedagogy... Just like the police learn from each other about what's working best, so do teachers and principals. M-Stat allows them to do it. It's all about sharing knowledge and experience. And it's about transferring that know-how, those proven practices, from one professional to another, from one school to another. It's about doing what works best to improve learning for kids...It's all about pedagogy."

Pedagogy. Good word.

M-Stat certainly brought about greater accountability and better results. It did this by creating a structured way for the whole enterprise of teaching in Montgomery County Public Schools to better and more quickly understand what worked. The district had built a new way of working, of collaborating, of diagnosing problems, and of making quicker adjustments and changes to address the learning needs of every student.

And it worked. In secondary schools, the achievement gap was greatly narrowed according to national measures like AP scores, Scholastic Assessment Test (SAT) scores, and high school graduation rates. It

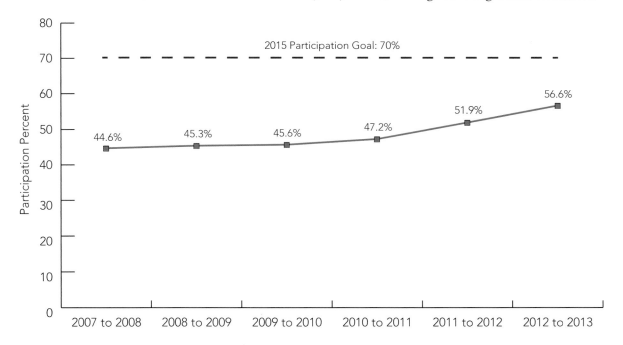

Increasing the numbers of our students who receive school breakfast—especially in our poorest and hungriest neighborhoods—became a big part not only of hitting our strategic goal for education, but also for hitting our strategic goal to eradicate childhood hunger in Maryland by 2015. During the school year ending in 2013,16,415 more students than the year before received a breakfast each day. The overall participation rate increased to 56.6 percent by 2013 compared to 51.9 percent in 2012, well on the way to our goal of 70 percent participation by 2015.

Jerry Weast on M-Stat

A Great County Schools Superintendent in His Own Words

Initially it was my job as leader of Montgomery County Public Schools (MCPS) to build a community consensus about the need for fundamental change. Through GIS mapping, we were able to geographically illustrate regions of poverty, high mobility, and low academic performance to build understanding of the downward spiral of educational outcomes that was the long-term trend for far too many children in the "Red Zone" of our county's map. Within the school system, we embarked on a process of continuous improvement. Exposing performance data not only forced an acknowledgment that the excellent public schools in the county were not serving all the children equally well, but also brought the issue of race squarely to the table.

After we defined the problem, we began to come to consensus about the goals we sought to achieve. Broadly stated, school system leaders agreed to pursue the interrelated goals of excellence—developing a curriculum that would ensure career and college-ready graduates, as well as equity. Working collaboratively, we moved into a phase that caused us to identify existing conditions and evaluate our efforts to determine if our resources were being used to best effect to improve outcomes.

Accelerating progress—engagement of staff, students, and families—required effective communication and better information sharing. But with the flood of data available, broader sharing was needed to get beyond anecdotal information and help people understand that correlation is not necessarily causation, but a piece of a bigger picture.

Modeling after CompStat, we developed M-Stat (short for MCPS Stat). Through M-Stat, data were developed around challenges, such as improving student readiness for success in Algebra 1 by eighth grade. Instructional leaders (principals and teachers) where success had been achieved would lead meetings of

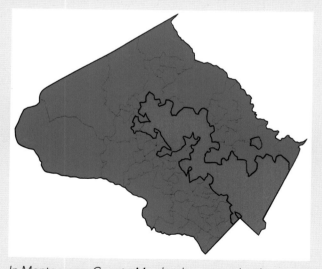

In Montgomery County, Maryland, county schools superintendent Jerry Weast had the guts to show us—on a map—the Green Zone, where students were achieving at above average levels, and the Red Zone, where students were performing at below average. When everybody knows, then everybody knows, and together we can do something to change the reality.

peers from other schools to share the processes and the pathways that were effective in that work. These proved to be highly successful in addressing the attainment of objectives. Creating a culture of continuous improvement requires effective leadership, but most of all, it demands collaboration if it is to succeed.

The Seven Keys to College Readiness—developed by back-mapping the academic histories of MCPS graduates who were successful in college—provide a pathway for students to follow that will increase their chances of being ready for and successful in college: advanced reading in grades K-2; advanced reading MSA in grades 3; advanced math in grade 5; Algebra 1 by grade 8, "C" or higher; Algebra 2 by grade 11, "C" or higher; 3 on the AP exam, 4 on the IB exam; and 1650 on the SAT, 24 on the ACT.

But results are what matter. Our students set district and national records on the SAT and ACT. In 2012 MCPS graduates scored an average 1651 on the SAT, 153 points above the national average. MCPS students and, notably, Black and Hispanic students, passed AP exams with a score of 3 or better at nearly three times the national rate. MCPS earned the highest graduation rate for African American male students among the nation's large school districts, and the highest high school graduation rate overall among large districts for four consecutive years. In primary grades the achievement gap in reading and math between Black and Hispanic students and their White and Asian peers all but disappeared. These gains occurred as the non-English speaking student population more than doubled and the school system was serving a growing number of children from families facing socio-economic challenges.

Beyond the classroom, MCPS improved school system operations, because a school system is more than students and teachers—it is buses, buildings, lunches and breakfasts, and dozens of other responsibilities that all must interconnect to make a school system great. Performance managment improves all of it. And MCPS received the Malcolm Baldrige National Quality Award for performance excellence in 2010. Empowering others to be decision makers—a responsibility that requires each person to gather information and advice from others before reaching a conclusion—pays dividends to the organization exponentially.

I do not believe there is a "big idea" lurking on the horizon that will transform public education to make it equitable and challenging for all children. Rather, there are a series of unglamorous but foundational changes that can be undertaken in any school district: review performance data so you know where you are; decide collaboratively what you want your students to achieve; evaluate the effectiveness of programs, technologies, and resource allocations; get rid of what doesn't work or is misaligned; and reallocate resources to support your goals. Repeat the process outlined above.

—Jerry Weast. Adapted from a chapter titled "Confronting the Achievement Gap: A District-Level Perspective" in *Improving the Odds for America's Children*, a compilation book by Harvard Education Press, 2014.

was also narrowed on interim measures like completion of Algebra 1 by eighth grade.

In a terrific book, *Leading for Equity*, by Stacey M. Childress, Denis P. Doyle, and David A. Thomas, a reflective Jerry Weast said, "I thought I would enter the change process through the culture door and then engage everyone in creating systems and structures that would support the culture. But I couldn't get traction, so we started to build the systems anyway, and it seemed the culture started to shift as people saw that the changes worked for kids."

Ending Childhood Hunger

"Solving poverty is complicated; feeding a child is simple." These are the words of Bill Shore, the tireless founder and leader of the nonprofit organization, Share Our Strength.

Children can't learn if they aren't fed.

In fall 2008, we set a strategic goal of eradicating childhood hunger by 2015. We were the first state in the union to set out to accomplish this, and many other states, led by a mix of Republican and Democratic governors, soon followed Maryland's lead.

Governing for Results: Ending Childhood Hunger

In 2008, we set a goal to eradicate childhood hunger in the State of Maryland by 2015. We measured our overall progress by the number of children eligible for free or reduced-price lunches who were eating a free or reduced-price breakfast at school every day.

Some of the things we accomplished in Maryland include:

- Through the 2013-2014 school year, more than 60 percent of children eating lunch in schools were also eating breakfast, which represented a 37 percent increase compared with the 2007-2008 school year.
- Through our efforts, Supplemental Nutrition Assistance Program (SNAP) participation increased for children under the age of eighteen from 6 percent in June 2009 to 9 percent through September 2014. In addition, in October 2014, the USDA announced that Maryland would be the recipient of a $3.4 million bonus incentive payment in recognition of the high percentage of eligible low-income families participating in the Food Supplement Program (FSP). Maryland's Program Access Index, USDA's measure of participation in FSP, was the highest of any state in the country.
- Data from Benefits Data Trust indicates that 173,876 unique households were reached; 37,899 of those households were educated and screened for FSP, 13,845 applications were submitted on behalf of low-income seniors and families with children, and 9,230 households enrolled in SNAP.

A StateStat dashboard showing progress across three of the six leading actions toward ending childhood hunger in Maryland by 2015 (as of July 1, 2014).

We identified five big strategies with goals to eradicate childhood hunger in our state:
- Ensure that all eligible families with children have access to supplemental food assistance.
- Provide all children in Maryland with access to a healthy breakfast.
- Expand the reach of summer meals programs for youth by serving one million addition meals.
- Expand access to nutritious food for pregnant women, new mothers, children, and youth.
- Enhance working families' economic security through expanded utilization of the Earned Income Tax Credit.

Although we had to expend some additional effort to coordinate and organize a web of caring people into concerted action, the food programs that allowed us to work toward eradicating childhood hunger were almost all funded through our federal government. Our challenge was to connect those programs with the hungry kids who needed food.

The programs that served our kids included:
- School Breakfast Program
- At-risk Afterschool Meals Program
- Food Supplement Program (FSP)/Supplemental Nutrition Assistance Program (SNAP)
- Women, Infants, and Children (WIC)
- Summer Food Service Program

But accessing these programs, doing the outreach, marketing the program, convincing principals and teachers to serve breakfast in the classroom as surely as reciting the Pledge of Allegiance—all of these were the leading actions that drove us to our goal. We didn't get all the way there by 2015, but we made a lot of progress toward getting there. And the well-being of our kids and their improved performance showed it.

There are a few individuals who deserve special thanks for their efforts in helping work toward meeting this goal: Anne Sheridan, executive director of the Governor's Office for Children; Lillian Lowery, state superintendent of schools; and Tom Vilsack, secretary of the US Department of Agriculture.

Lessons Learned: Achievement Gaps and the Learning Environment

Setting goals and tracking progress is not enough. But in a big complex endeavor like improving the public education system of a state, it is an essential foundation to more impactful and targeted local action. And yes, certainly, there are some local school districts with great leaders who don't need to be told what the larger statewide goal is, and there are a few gifted leaders who don't have much to learn from colleagues. But those genius talents are few and far between, and there is a lot the rest of us could learn from them.

Collaborative leadership is all about learning from one another. Learning from each other's experience and insights. Learning from those who work for us, and beside us. Some of the most important lessons I learned were those shared with me by my oldest daughter, who taught for three years in Baltimore City public schools with Teach for America.

The best curriculum and most highly dedicated teachers in the world will have a difficult time teaching students who are hungry, cold, or trying to learn in a classroom that has no windows and not enough oxygen. Improving the learning environment and the well-being of kids are critically important to the mission.

Understanding and improving the learning environment is essential for closing achievement gaps between children from poor families and children from wealthier ones. There are tangible things we can do to measure and improve that learning environment. There are tangible things we can do to close achievement gaps. These places are not unknown to us. They exist on all our maps. And in many places across America—especially poor places—we have yet to effectively tackle the challenge.

Education is not a gift. Education is a necessity for an inclusive economy. It is a right, regardless of economic class, race, and religion.

We know what works; we just need to do more of it.

"How many of you believe this job is really cool and really easy...?!"

Learn & Explore

School Locator
This app can be used to locate elementary and secondary schools, as well as to view school districts and attendance zone boundaries.

School Proficiency Index
This policy map presents data about the performance of fourth-grade students on state exams to describe which neighborhoods have high-performing elementary schools nearby and which are near lower-performing elementary schools.

Fight Child Poverty with Demographic Analysis
In this hands-on GIS lesson, you'll learn how to direct funding toward resources to help at-risk kids in Detroit.

For links to these and other examples, exercises, and resources, visit SmarterGovernment.com and click chapter 8.

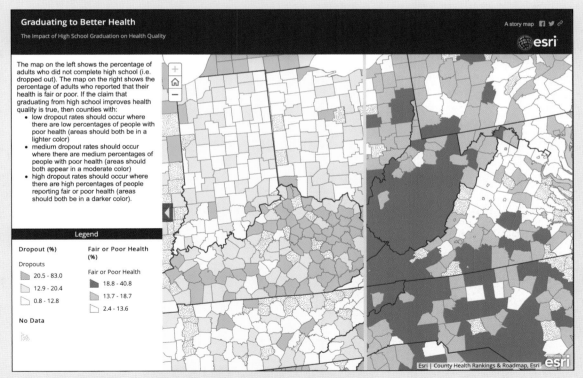

Graduating to Better Health: This story map lets you explore the correlation of high school graduation rates and health quality across the US.

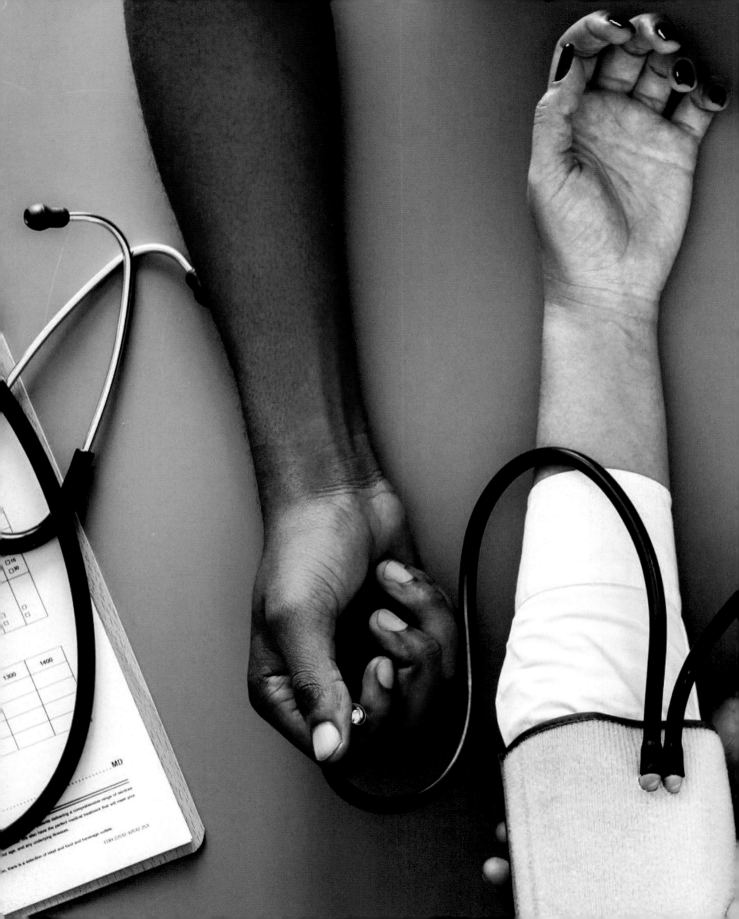

9

Improving Health and Well-Being

A new story is emerging all around the world. It is not a story of growth for the sake of growth. It is a story of greater health and well-being for all.

When it comes to eradicating disease, poverty, hunger, infant mortality, or other threats to human life, the question is always the same: Do we know where it is happening, and are we doing something about it? Are we doing something that works to save lives?

Transforming Community Health

If crime is the problem, you use the common platform of GIS to put the dots on the map where crime is happening and deploy police to where the crime is occurring to drive it down. And if hungry children can't learn in school, you use the common platform of GIS to plot where those children go to school, and you implement the routine of breakfast in the classrooms of those schools.

But I have found one group of professionals—more than any other—takes to this new way of governing like ducks to water. And that group is public health professionals. Epidemiology is the study of health and disease. More particularly, it is the study and analysis of the *distribution* of the *causes* of health and disease conditions in defined populations. When it comes down to it, *epidemiology is the science of who, what, when, where, and why*.

Knowing who, what, when, where, and why leads to the right questions about what we should do about it. Asking the right questions leads to the right array of life-saving actions.

If you knew that every year the vast majority of children poisoned by lead were poisoned in a handful of zip codes in your city or state, what would you do about it?

If drug addiction were the leading cause of death, disease, and property crime in your state, how and where would you go about treating it?

If incidents of infant mortality were spiking in your state, how, when, and where would you go about doing something to save those young lives?

Dusk over Federal Hill Park in South Baltimore.

The History of Using Maps and Data to Study Public Health Problems

A data- and performance-driven approach to understanding public health problems and identifying potential solutions dates to the time of John Snow—the middle of the nineteenth century, in London, England. In 1854, a terrible cholera outbreak ravaged the city. Snow, a local physician, decided to map where the people who suffered from the dreaded disease lived. He noticed a pattern—a concentration of cases around a water pump on Broad Street. Snow saw that the farther away from the pump he looked, the fewer and fewer cases of cholera.

Snow didn't have all the information he wanted—he suspected but could not confirm that cholera was a bacterial disease. He wasn't even sure the route of transmission was water. (Many people at the time blamed "miasma," a somewhat mysterious toxic vapor.) But he had good sense, and he had a map. And armed with data and the map, he convinced the local authorities to remove the handle of the pump, bringing an end to the cholera outbreak.

Snow is considered the founder of modern epidemiology, but he also pioneered the use of the data and the map to drive policy. In Baltimore and in Maryland, 150 years later, we followed his footsteps.

LeadStat: Reducing Childhood Lead Poisoning

For decades, children were being poisoned by ingesting lead paint dust and chips in some of Baltimore's oldest and most dilapidated properties. The effects were well documented—from behavior and learning problems to serious neurological damage and even a risk of death in cases of severe poisoning.

It was an outrage without outrage.

Lawyers were always running television advertisements soliciting parents whose children had been poisoned by lead. Unlike the city's lawyers, the private bar had no problem finding the courthouse. In the ten years before I took office, the city had not taken a single landlord to court to force them to clean up a lead-contaminated property.

City government seemed convinced the problem was too overwhelming to control or to address. The sentiment seemed to be, "Children get poisoned by lead in poor neighborhoods. That's Baltimore; there isn't much we can do about it. We don't want to drive out landlords and affordable rental housing."

In truth, there was no excuse for abandoning the children who were most at risk. So, we decided to tackle the problem head-on. It would require collaboration among a wide circle of city agencies and private nonprofits; a new city council ordinance; the cooperation of pediatricians, lawyers, and scores of private contractors; and a mix of federal, state, and municipal funding.

On a scale of one to ten, municipal difficulty would rate a six. We began to map and to measure.

We created a central registry of lead-contaminated properties and put the health commissioner in charge of keeping it current and accurate. We mapped the properties and started overlaying every reported case of a child testing positive for elevated lead levels. We found the cases were concentrated in two of the poorest parts of our city—neighborhoods where the housing was some the oldest.

We created LeadStat—an iterative process for measuring and mapping our progress toward eradication of lead poisoning. After some consideration, we put Health Commissioner Peter Beilenson in charge, with the deputy housing commissioner in a closely supportive role overseeing code enforcement. The Health Department would provide education and case management to all families with lead-exposed children, and Housing would use the courts to abate lead hazards in their homes. We recruited pro bono help from law firms to drive these enforcement actions through the courts.

We were relentless, and did not let up. We tracked this collaborative effort centrally, using the new lead registry to define the problems, using the map to create a common platform for Health and Housing to

Governing for Results: LeadStat

Many homes in Baltimore built prior to 1978 contain lead-based paint. Lead exposure can be a significant health risk, especially for young children under six years of age. Childhood lead poisoning can cause learning and behavioral problems, as well as more serious health consequences.

In Baltimore, we identified childhood lead poisoning prevention as a priority public health issue in 2000 and instituted LeadStat to measure and manage performance of our lead abatement efforts. Some of the things we accomplished with LeadStat include:

- We initially cleared cross-jurisdictional barriers and cross-trained health and housing inspectors to streamline enforcement.
- In the first year, lead was abated in 127 homes across the city.
- Fewer and fewer kids were poisoned and damaged by lead with every passing week. Between 2001 and 2003, we saw a 61 percent decrease in dangerously high blood lead levels in children six years old and younger. By 2007, this had decreased by 65 percent.
- Between 2000 and 2007, several hundred cases were filed against property owners in violation with the city's lead regulations.

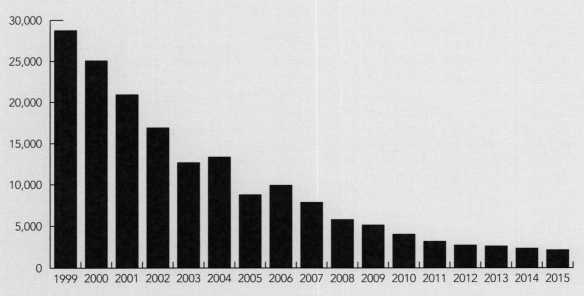

The number of children in Maryland who had lead levels of 5 micrograms or higher of lead per deciliter of blood was driven down dramatically between 1999 and 2015 thanks to data, the map, and the method of relentless follow-up through LeadStat. Among the leading actions—universal testing of toddlers, more rapid inspection and enforcement, and targeted dollars to help with timely remediation.

coordinate their efforts, and using regular LeadStat meetings for problem-solving conversations.

We pushed legislation through the city council that required screening for all one-year-old and two-year-old children.

We focused housing inspection efforts on homes where children with high lead levels in their bloodstreams had been identified. When our housing inspectors became overwhelmed, we trained firefighters to inspect homes for lead hazards, as well. When we learned that people were more likely to open their doors for firefighters than for city employees in suits, we teamed the suits up with firefighters to gain quicker access to vulnerable homes.

We joined forces with a couple of great nonprofits to increase our outreach, education, and remediation efforts. We went after a grant from the US Conference of Mayors—funded by DuPont, a responsible paint company—that allowed us to distribute do-it-yourself home testing kits to every expectant or new mother in the city, reaching more than 9,000 moms each year.

We created a primary prevention initiative to provide lead-safety training to home health care and other workers who served pregnant and postpartum high-risk, low-income women. When we identified pregnant women living in homes with lead risks, we followed up with education and supplies that enabled families to reduce lead and other health hazards before babies coming home from the hospital could be exposed.

We started doing a better job of accessing federal and state grants and loans to help pay for stabilizing and removing lead-contaminated components of a home, like doors and windows. This money helped us do risk assessments, hire contractors, and—when necessary—temporarily relocate families while the work was being done.

When lead was discovered in the drinking fountains of schools, our health commissioner shut off the water—many years before leaders in other cities did the same.

The harming of so many young lives was no longer something that "just happened" in poor neighborhoods.

If you think you can or think you can't, you are probably right.

DrugStat: Reducing Drug Addiction

The year I took office as mayor, the conventional wisdom was that Baltimore had 60,000 drug addicts—roughly one in ten of our citizens. But rather than shocking us into action, that sad claim had numbed us into a coma. Baltimore had not opened a new inpatient drug treatment center in nearly thirty years.

All the smart people knew there was nothing we could do about drug addiction until we had treatment on demand. While other community causes—museums, aquariums, symphony halls—were well-versed in the strategy of incremental progress, we believed that drug treatment had to be all or nothing.

Frustrated by the tail-chasing, I asked our health commissioner: "How much more treatment would it take for us to achieve the biggest reduction in overdose deaths in the country?"

"We need treatment on demand," he responded.

"Yes, I know. We need great schools on demand, good jobs on demand, decent housing on demand, crime-free neighborhoods, and loving homes for every child on demand. But we get paid to make progress and save individual lives. How much more treatment," I asked again, "will it take for us to make nation-leading progress in reducing drug overdose deaths?"

It was a different question—a question that did not allow for a default to the impossible. We might not be able to secure funding for drug treatment on demand, but we could double or triple its availability.

Tackling Drug Addiction and Overdoses in Baltimore

Baltimore has a long history with heroin, dating back decades. In 1999, 312 city residents died of heroin-related overdoses, adding to the city's misery. But there was reason for hope. Evidence was emerging from a series of studies in Baltimore showing that treatment services helped people escape the grip of addiction.

As CitiStat changed people's expectations about city services, Health Commissioner Dr. Peter Beilenson established DrugStat to improve the quality of new treatment options across the city. In 2002, researchers at Johns Hopkins University, the University of Maryland, and Morgan State University released the results of the Baltimore Drug and Alcohol Treatment Outcomes Study, "the largest and most rigorously conducted drug treatment outcomes study that focuses on a single city." The report found that treatment was associated with large reductions in heroin use, cocaine use, drinking to intoxication, criminal activity, arrests, and HIV risk behavior—and these reductions were sustained through twelve months of follow-up. "With this data," wrote Dr. Beilenson, "comes a public health responsibility to make 'treatment on demand' a reality."

The first major steps toward "treatment on demand" had come when Mayor O'Malley won more than $30 million in extra funding each year for addiction treatment from the governor and state legislature. What was previously a trickle of financial resources became a steady stream. Baltimore's treatment system grew to accommodate thousands of people in need, but long waitlists remained.

By 2005, a new tool had become available to fight heroin addiction: the medication buprenorphine. Like methadone, buprenorphine is a long-acting opioid that can address the craving of addiction without causing euphoria, allowing people to regain control over their lives. Unlike methadone, buprenorphine had the unique chemical property of antagonizing itself at high doses, making overdose less likely. As a result, Congress allowed regular doctors in practice, after taking a training course, to prescribe buprenorphine—creating a path to significant expansions in access to care.

"Have you started the buprenorphine initiative yet?" Mayor O'Malley buzzed his new health commissioner on his first day of work in December 2005. Within several months, Baltimore became the first city in the country to pay for the training of every doctor interested in prescribing buprenorphine. The city established a referral system that started patients in an intensive addiction treatment program, gave them all case managers, and referred them to primary care doctors for ongoing care. When a patient relapsed, the primary care doctor could call the case manager and refer to the addiction treatment center. It was "hub and spoke" before "hub and spoke" became a name for this kind of treatment system. Meanwhile, the regular, data-driven DrugStat meetings allowed for focused oversight, shared learning, and updating of protocols.

And then, a policy breakthrough. After winning statewide election in 2006, Governor O'Malley and John Colmers, his secretary of Health and Mental Hygiene, got to work expanding access to health insurance. An early goal was paying for addiction treatment through the Medicaid program, which would bring in a 50 percent federal match, doubling the reach of state funds. John Colmers saw his opportunity in the Primary Adult Care Program, a small effort started by the previous governor that covered primary care services and pharmacy benefits for low-income, otherwise uninsured adults. Even before expanding the full Medicaid program, and years before the Affordable Care Act, Maryland added addiction treatment benefits to the Primary Adult Care Program. This one step made every person suffering from an addiction in Baltimore eligible for insurance that covered outpatient addiction treatment, including treatment with methadone or buprenorphine. One result was the steady elimination of waiting lists in many parts of Baltimore, something previously considered an unattainable goal.

Another result? A gradual fall in the number of overdose deaths from heroin. From a peak of 313 deaths in 1999, the toll fell to 76 in 2011, a decline of more than 75 percent. A study in the *American Journal of Public Health* found "a statistically significant inverse relationship between heroin overdose deaths and patients treated with buprenorphine" with deaths falling, on average, by 37 percent after buprenorphine became available. With the passage of time, a change in administrations, the drifting of priorities, and the availability of cheap heroin tainted with fentanyl, Baltimore and many other areas on the East Coast and in the Midwest faced a new, far more lethal wave of heroin overdoses. Some would throw up their hands and say that there was nothing that could be done.

Those aware of Baltimore's historic success knew otherwise. Indeed, in 2018, those cities and states who responded to the challenge by investing in treatment systems—some borrowing from protocols developed a decade earlier in Baltimore—are seeing progress. Taking a page from Baltimore's playbook, Rhode Island Governor Gina Raimondo invested in centers of excellence for buprenorphine treatment, encouraged primary care doctors to treat patients, and took the additional steps of encouraging hospitals to provide peer counselors to overdose victims in the emergency room and of offering treatment to everyone in the state jail and prison. Overdose deaths among those leaving detention were reduced by 60 percent, and for the first time since fentanyl-laced heroin hit the United States, Rhode Island saw a decline in overdose deaths in 2017.

—Dr. Joshua M. Sharfstein, professor of the practice in Health Policy and Management at the Johns Hopkins Bloomberg School of Public Health and former Baltimore health commissioner

On the one-to-ten scale of municipal difficulty, this would be a ten. It would require collaboration, political persuasion, and performance measurement around one of the most challenging variables of all: a human being in the throes of addiction. Our Health Department, our hospitals, and our courts would need to collaborate. We would need to keep existing partners on board while bringing new foundations to the table. We would need to fight back against the stigma that saw drug addiction primarily as a moral failure, as a matter for the police and the courts rather than as a matter for treatment and recovery. We would need to develop, for the first time, a way to measure the performance of various drug treatment providers and reward the most effective.

And right up front we would have to find some way to reverse our chronic underfunding of the problem. The policing effort absorbed every dime of the local dollars we had. But I did have the ability to set our legislative agenda in Annapolis. We made drug treatment the single biggest budget request we took to the state when I first took office as mayor.

Only the governor had the authority to propose budget allocations—the legislature could only cut. But with Pete Rawlings, chair of House Appropriations, and Senator Barbara Hoffman, chair of the Senate Budget and Taxation Committee, the Baltimore delegation had influence greater than its numbers. Whether my delegation would follow their mayor and fight for his requested priority was another matter.

We asked for $25 million for drug treatment. The most that had ever been allocated statewide was $10 million, with $8 million going to Baltimore.

We had to persuade Democratic Governor Parris Glendening to fund—for the first time—a huge increase in treatment for people who were largely uninsured and unemployed, who had no organized constituency, and who generally did not vote. It was hard for many in our delegation to picture a ribbon cutting. We had quite a few conversations about the downside of using all our political capital on such a single big request.

But when the talking was done, the venerable dean of our delegation, Senator Clarence Blount, called the question. "Everyone has had their say, but we are one city and we have one mayor. And the mayor tells us that this funding is critical to our city."

God bless Senator Blount—my delegation backed me up.

Initially, Governor Glendening submitted a budget that only increased our funding to $11 million. But by the end of that first legislative session, our delegation and its powerful budget chairs had persuaded him to increase it to $18 million. We had effectively doubled our funding. And over the next few years, thanks to Governor Glendening, funding for drug treatment in Baltimore would go up to $51 million a year—more than six times what it had been in 1999.

With greater input secured, we could now focus on the output—saving more lives.

Not all drug treatment programs worked well, and some worked better than others. We would reward programs that delivered the best results. DrugStat became our performance-measured collaborative process. Working with treatment professionals, we developed a more holistic set of indicators of success. It would be nice if healing someone from an addiction were as simple as a forty-eight-hour stay in detox or even a twenty-eight-day stay in an inpatient facility. It's not.

We tracked frequency of urinalysis, relapses, days clean, and percentage of participants clean by thirty days, by sixty days, and by ninety days. But we also tracked the number of people in a given program who had a stable address by a date certain, who had gone on three job interviews by sixty days in recovery, and who had found a job by ninety days in recovery. Our outcome measures also included crime, which, we noticed, began to drop as treatment became more available.

One of the most critical problems we faced in ramping up was a dearth of inpatient drug treatment

Governing for Results: DrugStat

In 2001, one out of every ten Baltimore residents was addicted to drugs. We introduced DrugStat to help us expand, measure, monitor, and increase the performance of our drug treatment programs. Some of the things we accomplished with DrugStat include:

- Between 2001 and 2002, we increased the six-month retention rates in our methadone programs by 16 percent.
- In 2002, Baltimore achieved the largest two-year decline in overdose and other drug-related hospital admissions (a 36 percent decline for heroin overdose cases, and an 18 percent overall decrease).
- From 1997 to 2005, the number of residents who participated in publicly funded treatment programs increased from 18,449 to 28,672. And we finally had a way to hold treatment programs accountable. We defunded programs that did not perform.
- Based on these results, the State of Maryland increased funding for our drug treatment program by an additional $25 million over a four-year period.
- By 2005, drug-related deaths in Baltimore were driven down to their lowest point in more than ten years. They would continue to be driven down for another five years before the recent spike in heroin and fentanyl deaths.

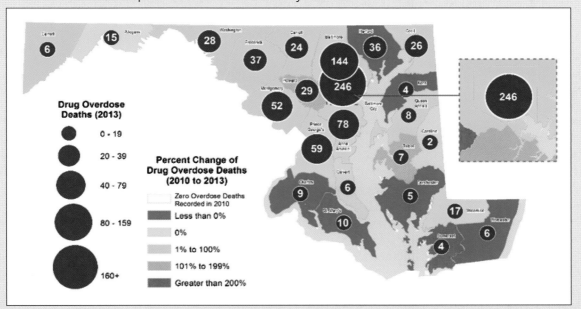

Maryland overdose deaths in 2013 and change from 2010 to 2013 by county. For years we drove down overdose deaths through a combination of different and better coordinated actions whose effectiveness was measured and better coordinated through DrugStat. For many years it seemed making drug treatment more widely available in conjunction with mental health treatment, housing, and employment services was the answer. Then the opioid epidemic hit. As our problems change, so must our tactics and strategies.

beds to help people who needed a shot of stability before returning to community-based treatment. So, the Harry and Jeanette Weinberg Foundation, the Abell Foundation, and the France-Merrick Foundation joined forces with the Open Society Institute to bring a desperately needed bricks-and-mortar project into being.

Finding a place to build it was every bit as difficult as securing the funding. Everyone likes the idea of more drug treatment, but few people will readily accept a treatment center in their neighborhood. After much discussion and city council debate, our Park Heights neighborhood stepped up to the plate. I have attended many openings—of new libraries, schools, college buildings, bridges, and commuter rail stations. None has filled me with such deep satisfaction or pride in the goodness of my neighbors as the opening in September 2006 of the 120-bed Gaudenzia inpatient drug treatment center in Park Heights.

In the meantime—because perfect should not be the enemy of good—we worked with the funding and tools we had. Our primary indicators of success were drug overdose deaths and "drug related hospital emergency room admissions"—a somewhat broad and subjective category but one that all our hospitals tracked. As we progressed, a third indicator emerged—property crimes—and these were tracked routinely by the police. It takes a lot of petty thefts and burglaries to feed a thousand-dollar-a-day habit.

Applying the same Stat techniques Jack Maple had pioneered to help fight crime in New York City, we began to battle our worst-in-the-nation addiction problem in Baltimore.

Even without treatment on demand, we drove down overdose deaths by a third—from 321 in 1999 to 214 in 2005. That was more than one hundred lives saved a year. By 2010, that number would be driven down by 75 percent compared to 1999.

In my last year in office as mayor, we began to expand the use of buprenorphine treatment for opioid use disorder and offered training to doctors throughout our city. This dramatically expanded access to treatment and helped save even more lives.

Reducing Infant Mortality

It is hard to imagine a more unfathomable sorrow than the grief of burying one's own child.

Infant mortality reached its height in Maryland in 2004 with 632 infant deaths, and my state administration set out to tackle this scourge. In 2007, we set "reducing infant mortality by 10 percent" as one of our state's sixteen strategic goals. Thanks to the leadership of Secretary Joshua M. Sharfstein at the Department of Health and Mental Hygiene (DHMH), Secretary Ted Dallas at the Department of Human Services (DHS), and many very dedicated public health workers, by 2015 we had succeeded in reducing Maryland's infant mortality rate by 17 percent.

By deploying our outreach, interventions, and support services for expectant moms to the "hot spots" where infant mortality rates were highest in our state, we saved a lot of babies' lives. Faster processing of federal benefits for women, infants, and children also helped save lives. The extension of health care coverage to seven hundred thousand Marylanders—who had no coverage before—likewise helped us achieve these record reductions.

Many complex factors contribute to infant mortality. If women are healthy at the time of conception, the outcome of the pregnancy is much more likely to be positive. So, access to comprehensive women's health care was prioritized by DHMH—this meant knocking on more doors and doing more outreach.

Beginning in 2011, the number of family planning clinics was expanded from three target jurisdictions to eight: Baltimore, Caroline, Charles, Dorchester, Prince George's, Somerset, and Wicomico Counties, as well as Baltimore City. The number of women receiving comprehensive women's health services in

Overall Progress	Total Infant Mortality Rate (Deaths per 1,000 Live Births)	African American Infant Mortality Rate (Deaths per 1,000 Live Births)	White Infant Mortality Rate (Deaths per 1,000 Live Births)
0% 10% 21.3% Decrease in Maryland's Infant Mortality Rate	2012 **6.3** Deaths/1,000 Live Births 2007 **8.0** Deaths/1,000 Live Births	2012 **10.3** Deaths/1,000 Live Births 2007 **14.0** Deaths/1,000 Live Births	2012 **4.1** Deaths/1,000 Live Births 2007 **4.6** Deaths/1,000 Live Births
ON TRACK ▲	21.3% DECREASE ▲	26.4% DECREASE ▲	10.9% DECREASE ▲

A StateStat dashboard showing progress towards our statewide strategic goal of reducing infant mortality by 10 percent by 2017 (as of July 2, 2014). There were a finite number of expectant moms in vulnerable conditions who we learned to do a much better job of reaching and helping with pre- and post-natal care, support, and nutrition.

targeted jurisdictions increased, in large part due to the expansion of outreach efforts. In 2013, more than 23,000 women received comprehensive women's health services in these eight targeted jurisdictions across the state. The monthly average number of women who received services in target jurisdictions increased from 312 in 2011 to 1,084 in 2014 through September (excluding Prince George's County)—an increase of 247 percent. Expansion of the Medicaid Family Planning Program provided an additional 31,000 women who had no insurance coverage after their pregnancy with a limited benefits package that covered visits to a doctor, physical exams, and birth control, among other things.

Risk-targeted prenatal care was critical for preventing infant deaths. Part of DHMH's plan for reducing infant mortality included the implementation of Quick Start prenatal care programs in targeted jurisdictions. These programs ensured immediate access to care while Medicaid eligibility and the first prenatal care appointments were pending. Starting in 2011, DHMH expanded Quick Start to seven target jurisdictions, and the monthly average number of visits in targeted jurisdictions increased from 55 in 2011, to 142 in 2014.

In September 2013, the State of Maryland was awarded a $6.4 million competitive grant to expand evidence-based home visiting programs. Programs in five county jurisdictions and Baltimore City were expanded, as a result; by September 2014, nearly six hundred more families were enrolled.

Risk-appropriate follow-up is essential for mothers and infants with poor pregnancy outcomes, including pre-term and very low birth weight. To ensure follow-up occurred, DHMH implemented a standardized hospital postpartum discharge form to link high-risk mothers and infants with community services. DHMH also developed a model breastfeeding policy for all Maryland birthing hospitals. As of 2014, all thirty-two birthing hospitals had committed to adopting Maryland's breastfeeding policy recommendations or were working to become certified "baby friendly." We also reviewed and updated the Maryland Perinatal System standards to ensure that Maryland's standards were in line with guidelines and policies published by national professional organizations.

We also worked to prevent infant deaths that result from improper care and environmental risks. In 2012, working with partners in Baltimore City, DHMH distributed 3,000 *Sleep Safe* DVDs as part of the B'More for Healthy Babies "Sleep Safe" campaign. The Safe Sleep efforts continued through DHMH's website, a B'More for Healthy Babies website, print materials, and online videos. The Sleep Safe campaign video had more than 8,000 views on YouTube. A Spanish-language video and a video that targets fathers also are available on the B'More for Healthy Babies website. Also, in 2013, Maryland banned the sale of baby bumper pads statewide, preventing the introduction of this unnecessary risk from the homes of families in Maryland.

Although federal data was often delayed, Maryland had systems in place to capture and analyze data quickly, like data produced by the Child Fatality Review Team (CFRT). Established by statute in 1999, this multi-agency, multi-disciplinary team led local teams that were charged with reviewing unexpected child deaths, including homicide, suicide, and natural, accidental, and undetermined-cause deaths. Data and information gathered in 2012 was used by local CFRTs to develop recommendations and innovative approaches to prevent infant deaths related to sleep habits. The review process resulted in fourteen jurisdictions conducting safe sleep activities because of the CFRT's findings. Between 2007 and 2015, the number of child fatality review referrals declined by 39 percent, from 302 to 185.

After achieving the initial goal of reducing infant mortality by 10 percent by 2012, my administration reset the bar, striving for an additional 10 percent reduction by 2017. We also made substantial progress in reducing the infant mortality rate among African American moms, which reduced from 14 deaths per 1,000 live births in 2007 to 10.5 in 2013.

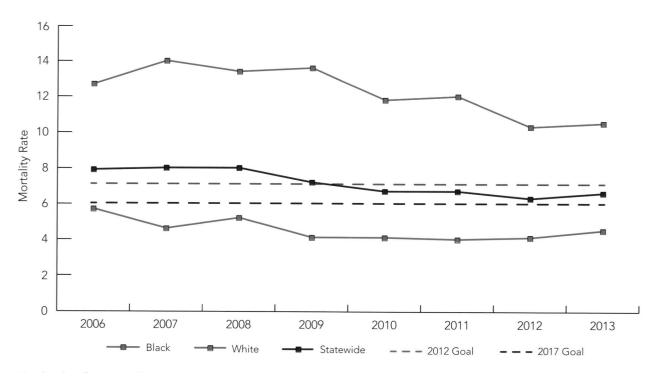

Maryland's infant mortality rate (measured in deaths per 1,000 live births). After achieving the initial goal of reducing infant mortality by 10 percent by 2012, we reset the bar to strive for an additional 10 percent reduction by 2017. The difference between a dream and a goal is a deadline.

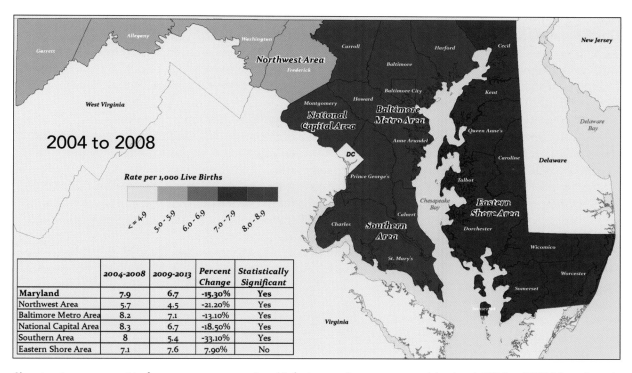

	2004-2008	2009-2013	Percent Change	Statistically Significant
Maryland	7.9	6.7	-15.30%	Yes
Northwest Area	5.7	4.5	-21.20%	Yes
Baltimore Metro Area	8.2	7.1	-13.10%	Yes
National Capital Area	8.3	6.7	-18.50%	Yes
Southern Area	8	5.4	-33.10%	Yes
Eastern Shore Area	7.1	7.6	7.90%	No

Showing improvement in five-year average regional infant mortality rates across Maryland, 2004 to 2008 (above), and 2009 to 2013 (below).

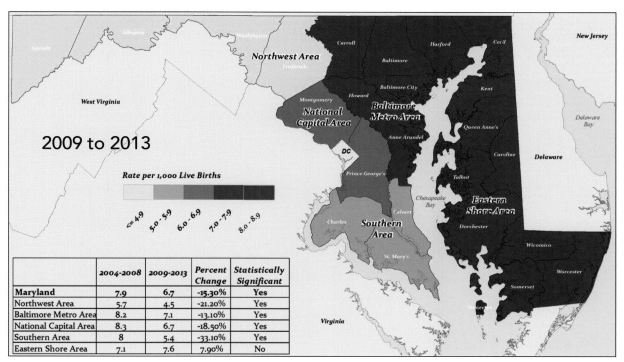

	2004-2008	2009-2013	Percent Change	Statistically Significant
Maryland	7.9	6.7	-15.30%	Yes
Northwest Area	5.7	4.5	-21.20%	Yes
Baltimore Metro Area	8.2	7.1	-13.10%	Yes
National Capital Area	8.3	6.7	-18.50%	Yes
Southern Area	8	5.4	-33.10%	Yes
Eastern Shore Area	7.1	7.6	7.90%	No

CRISP: A Common Platform for Health Information

Early in my administration in Maryland, we had no way to put individual health information on the common platform of GIS to better care for patients suffering from chronic conditions like congestive heart failure, diabetes, and asthma—the big drivers of avoidable hospital re-admissions, the biggest driver of unnecessary health care costs.

In 2007, we supported the creation of the Chesapeake Regional Information System for our Patients (CRISP). And given the privacy imperative, it was a complicated information technology undertaking that was accomplished methodically and successfully over eight years. Because it went well, you might never have read about it in the paper or seen it in the news.

Prior to this innovation, an individual's health data was fragmented across many electronic medical records or electronic health records (EHRs)—or, worse, just recorded on paper—and not easily shared among care givers. It's harder to help a patient stay well if you don't know about all the patient's other visits at other hospitals.

With the establishment and development of CRISP, doctors and nurses could securely access health data, when a patient consents, from across every hospital in the state. As of 2016, doctors and nurses were using the system more than 2,000 times every day to make more informed treatment and diagnosis decisions. Along the way, "CRISP" became a verb in Maryland. An emergency department physician might ask a colleague to check the system to find out key historical information about a patient who had arrived for care: "Can you CRISP the patient?"

CRISP could also securely notify a primary care doctor or care manager whenever one of their patients arrived at a hospital emergency room anywhere in the state. This started happening hundreds of times every single day, helping to ensure Maryland patients were getting the coordinated care they deserved and needed.

Inspired by the great work of the Camden Coalition, which pioneered the use of performance measurement for public health on a municipal level, we were able to enhance the CRISP technology and apply it statewide. For the first time, we could analyze the big picture of public health across Maryland.

A StateStat dashboard showing progress towards the statewide strategic goal of reducing preventable hospitalizations by 10 percent by the end of 2015 (as of July 2, 2014). Instead of paying hospitals like hotels (based on high occupancy), we started setting global budgets to make it more profitable for hospitals to keep patients well at home rather than sick and filling up the hospitals.

This map shows health-care spending around Baltimore in 2018, ranging from a high of $9,560 to $24,820 per household in the darkest blue areas, to a low of $0 to $1,410 per household in the lightest blue areas.

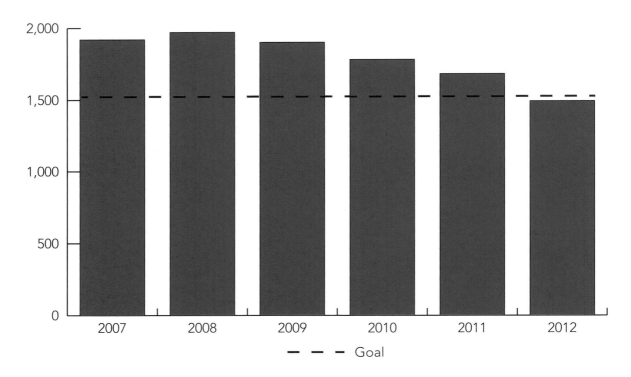

Preventable hospitalizations per 100,000 people in Maryland, 2007 to 2012. Today, Maryland is unique among the 50 US states, Maryland has a rate-setting commission that uses big data to align hospital reimbursements to incentivize the wellness of patients rather than the sickness of their symptoms.

We could identify clusters, or hot spots, of patients suffering from specific diseases. We could provide meaningful health data to eighteen public-private coalitions working on health improvement and well-being across the state. And we could empower community-level care managers with the same information to improve individual wellness at a neighborhood level.

The partnership of state agencies, CRISP, and health care providers that created this common platform for health had a powerful new tool for rapid analysis of health across the state. For the first time, they had the ability to generate detailed maps of health issues—all done with an underlying commitment to the privacy and security of individual health data.

Paying for Wellness

If you knew that avoidable hospital re-admissions were the biggest driver of unnecessary health care costs, how would you go about reducing them throughout a state?

In 2014, Maryland became the first state in the union to stop paying its hospitals like hotels; namely, on a fee for service based on how many beds they can fill. Under this old system—a system in use across the United States today—the more rooms a hospital fills with sick patients, the more profitable they are. Under this "All-Payer" or "Pay for Wellness" Reform, Maryland began to pay its hospitals a fixed annual rate based, in part, on past usage and performance. If the hospitals take actions or engage in practices that reduce avoidable hospital re-admissions, they get to keep a big portion of the resulting savings.

Governing for Results: Maryland's All-Payer System

In the early 1970s, the State of Maryland established an independent commission to control hospital prices through an all-payer rate-setting. In January 2014, Maryland, in cooperation with the Centers for Medicare & Medicaid Services, repurposed this all-payer system, moving it from a price-controlled, fee-for-service arrangement to one that established global budgets for hospital services. Under this arrangement, hospitals are guaranteed a predictable level of revenues from all payers—regardless of the volume of services provided. If hospitals can take actions to improve the well-being of their patients and thereby reduce avoidable hospital admissions, they get to keep the savings derived from keeping people well. This approach aligns a hospital's financial goals with the health of their communities, incentivizing the prevention of acute illness rather than focusing on admitting as many patients as possible for as many services as possible. Maryland made these and other accomplishments with the all-payer system:

- Hospital spending in Maryland was held to 1.53 percent growth per year from 2014 to 2016, well below the 3.58 percent rate of growth of Maryland's economy.
- Among the Medicare population, Maryland achieved a 19.9 percent reduction in hospital admissions, a corresponding 20.5 percent increase in emergency department use without admission, and a 9.4 percent decline in ambulatory care-sensitive admissions.
- In the first three years, hospital spending in the Medicare program alone was reduced by $586 million compared with growth matching the national trend.

HOW IT WORKS

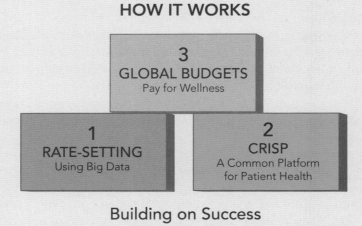

Building on Success

The three building blocks of Maryland's Pay for Wellness model.

The hypothesis of this reform was that if the bottom line of the hospital were guaranteed, administrators would have the freedom to redeploy effort to those actions that keep people from returning to the hospital. Their "reimbursement" and their profit for prevention and wellness activities comes from the savings they get to keep with their guaranteed annual budget.

In its first few years, the Maryland model has been so successful in reducing avoidable re-admissions and reducing hospital costs that it has saved Medicaid and Medicare hundreds of millions of dollars— and those savings to the federal government have been achieved ahead of schedule. Such was our success that the Obama Administration agreement with Maryland was renewed by the Trump Administration.

The mechanisms that allowed Maryland to make this change are human creations that any state with the political will could implement. One of those mechanisms is a rate-setting commission. Established by state law in the early 1970s, Maryland's rate-setting commission (the Health Services Cost Review Commission) has been setting the rate that hospitals are permitted to charge for certain services for going on almost fifty years now. Originally set-up to more equitably spread the cost of uncompensated care among hospitals, the institutionalized use of big data over the course of time gave Maryland the ability to model future utilization for every hospital based on past evidence.

The second mechanism was the ability to share individual patient health records across the entire system. This visibility allows public health professionals and primary care physicians along with hospitals to identify the "frequent fliers"—those patients whose poorly treated chronic conditions have them returning to the emergency room and the hospital time and time again.

Medical technicians in training at a career and technical education center in Baltimore.

What's been harder to pin down than the data and the map, is the method and the leadership. At the outset, hospital administrators are the leaders driving from the center of their individual hospital circles. But some administrators are better than others. Some are collaborative leaders; others engage in malicious obedience—doing the bare minimum to survive under the new rules while hoping the new system will fail. Most fall somewhere in the middle—between the leaders and the slackers. Where are the county health commissioners in this new system? Where is the state secretary of health? Are they passive observers or are they at the table? And whose job is it to convene people around that table?

To improve health outcomes for patients while reducing wasteful costs, a new cadence of accountability is needed. New circles of collaboration. New ways of sharing what works. And a repeated focus on the latest emerging truth.

The Maryland model is new. And so far, its success has exceeded expectations. But improving its agility, timeliness, and effectiveness will take persistent work. It will also require collaborative leadership, and higher uses of the data, the map, and the method of performance management.

Learn & Explore

HealthStat 2005
A sample HealthStat report from 2005, from the Baltimore City Health Department.

Heroin/Fentanyl Overdose Deaths
Will County, Illinois, heroin/fentanyl overdose deaths dashboard.

Fighting against the Opioid Epidemic
Direct Relief uses technology to distribute naloxone to combat the opioid epidemic.

For links to these and other examples, exercises, and resources, visit
SmarterGovernment.com and click chapter 9.

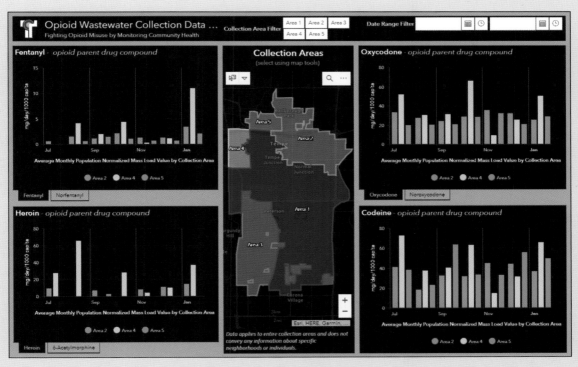

Data scientists at Arizona State University worked with the City of Tempe, Arizona, to analyze wastewater samples for traces of opioids to better understand the geography of addiction across the city.

10

Restoring Our Waters

Water is life. And for 350 years we watched as the life of our state—the waters of the Chesapeake Bay—declined in health year after year. But with GIS technology, more life-giving actions, and the measuring of performance against two-year milestones, we were able to reverse a 350-year decline.

The waters of the Chesapeake Bay are now getting healthier every year, not sicker.

Progress is a choice.

Water is life
From the dawn of the day
To the death of each night

All of the oceans
Rivers and streams
Filling our lives up
Enriching our dreams

For brother and sister
Water is life
Mni Wiconi
Water is life...

—Luka Bloom, "Water Is Life (Mni Wiconi)"

The Beauty of the Chesapeake Bay

The Chesapeake Bay is the largest estuary in the United States; its watershed covers the lands of six different states. It is a complex web of natural living systems—plant, animal, and water—all mutually dependent upon one another for life. Together, these systems form the living infrastructure—the green and blue infrastructure—that delivers critical, life-sustaining services that enrich and sustain the quality of life for millions of people and countless other living things.

In addition to its priceless environmental value, the Chesapeake Bay is also the source of jobs, income, food, and prosperity for countless thousands of residents in this generation and the next.

H. L. Mencken once wrote that Maryland is a mysterious and beautiful state. At the heart of that mystery and beauty is the Chesapeake Bay. For 350 years, so many people came to love the Bay so much that we nearly killed it.

The Need to Renew the Chesapeake Bay

The work of restoring the Chesapeake Bay is like walking up a downward-moving escalator—an escalator whose downward movement is being constantly accelerated by population growth, suburban sprawl, and land development.

By the 1970s, the health of its waters had become seriously degraded and threatened by the pace of human development on its shores and by the water pollution that we—often unknowingly—pumped into its rivers and streams. As beach closures and the collapse of rockfish and oyster populations made headlines, many of us began to ask ourselves a previously unimaginable question:

What would happen if the waters of the Chesapeake Bay were dead; devoid of life; incapable of maintaining their health; incapable of sustaining fish, plant, animal, or human life?

The declining health of the Chesapeake Bay kind of crept up on many of us. But the cause of the decline was not really a mystery. It was caused by us: by human activity on its surrounding lands. Building, paving, clear-cutting, fertilizing, and flushing: All these human activities on land pumped ever-increasing loads of nitrogen, phosphorus, and sedimentary pollution into the streams, rivers, and waters of the Chesapeake Bay.

At the same time, the native oyster population—once so plentiful that they served as natural subaquatic filters for the entirety of these waters—was decimated by over-harvesting and disease. Even the small feeder fish, called menhaden, were more recently fished to near depletion by industrial over-harvesting.

The combined result was more and more pollution with less and less of it filtered by forested stream buffers, marshes, natural shorelines, or native oysters.

A deadly cycle ensued.

With more nitrogen and phosphorus being carried into the Bay's waters, algae blooms fed by this pollution grew exponentially. When the algae blooms grow and then die, the decaying algae soaks up the natural oxygen in the Bay waters. Fish and other creatures literally suffocate, leave, or die of disease. Subaquatic Bay grasses get shut off from the sunlight, and then they wither and die. This deadly cycle was allowed to expand and repeat itself for years and years—350 years to be exact. One of the many measurable signs of this decline was a giant and expanding "dead zone" in the central stem of the Bay.

Were it not for the actions taken by Governor Harry Hughes during his administration from 1979 to 1987—and by other Democratic and Republican governors of Maryland since—the Bay would have died decades ago. We had the benefit of decades of work and science. And a healthier Bay was a dream we all shared. But a dream is not a plan.

As I once heard Senator Barbara Mikulski observe, "We have lots of programs for the Bay, but there's not much of a Bay program." We lacked meaningful benchmarks with deadlines. And we lacked an actionable plan that all could see and measure for themselves.

The Goal: The Healthier Bay Tipping Point

In 2007, we set a deadline for the strategic goal of restoring the health of the Chesapeake Bay. Our goal, *with a date*, was to reach the "Healthier Bay Tipping Point" by 2025. It wasn't exactly a "moon shot," but many of us thought it might be the last shot for the Bay. We defined the "Healthier Bay Tipping Point" as that point in time when each of her major riversheds were made a little bit *healthier* every year instead of always a little sicker. Getting to this point would require taking better actions to greater scale—a scale that could then join forces with the regenerative powers of nature to touch off cycles of healing.

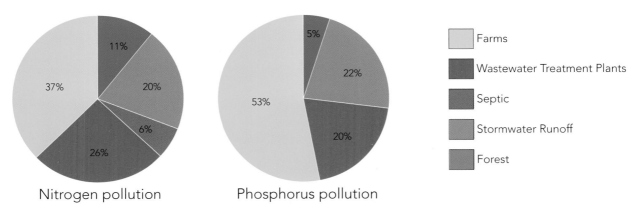

There are four major causes of nitrogen and phosphorus pollution flows into the Chesapeake Bay: discharge from wastewater treatment plants; fertilizer runoff from farms; leachate from housing developments served by septic systems; and stormwater runoff from roads, parking lots, and other hard-paved surfaces.

To meet this goal, our administration implemented (and in some cases strengthened) several actions—in fact, a suite of thirty-three *leading actions*, from planting more cover crops to upgrading wastewater treatment plants. The purpose of each of these actions was to reduce the volume of nitrogen, phosphorus, and sedimentary pollution that runs off into the Bay.

To map, measure, and drive progress to our goal, we created a new performance management system. We called it BayStat.

Having seen what CompStat did to improve public safety, and what CitiStat had done to improve the life of a city, I was eager to see if BayStat could restore the health of the Chesapeake Bay.

BayStat: The Data and the Map

Like its predecessors, CompStat and CitiStat, BayStat operated by way of a common platform—a GIS map of the Bay watershed within our state's borders. We divided this map into the ten main riversheds of the state, and as a practical matter, we could go down to a parcel level of granularity. Using this common platform, we could overlay a host of datasets and corresponding borders, from soil conservation districts, to stream buffers, to county and municipal borders.

What Is BayStat?

BayStat is a program created for the assessment, coordination, and reporting of restoration efforts for Maryland's portion of the Chesapeake Bay, as well as each of the ten major tributary basins.

The BayStat team included the governor; the director of StateStat; secretaries of the Maryland Departments of Agriculture, the Environment, Natural Resources, and Planning; scientists from the University of Maryland; other key staff of state agencies; the chief innovation officer; the budget director; and the governor's legal counsel. The BayStat team met monthly to assess progress and evaluate efforts.

To build public consensus and to drive progress with openness and transparency, BayStat shared information on the internet by way of easy-to-understand dashboards and graphs. Sharing this information with all was one way to better engage Maryland's citizens in the work of restoring the Chesapeake Bay.

Access to various Bay health-tracking tools provided members of the public with current information on the health of the Bay including real-time Chesapeake Bay water quality monitoring graphs, interactive maps showing pollution sources, and progress toward meeting strategic goals and milestones.

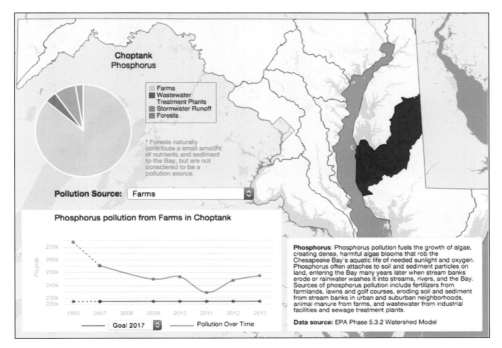

From the BayStat website: The primary causes of water pollution vary according to primary land uses from one rivershed to the next. Above, on the Eastern Shore, heavy agricultural uses dominate, so the load from fertilizer runoff is the leading cause of Bay damage. On the Western Shore, below, where housing developments on peninsulas are served primarily by septic systems, leachate and stormwater runoff are the leading causes of Bay damage.

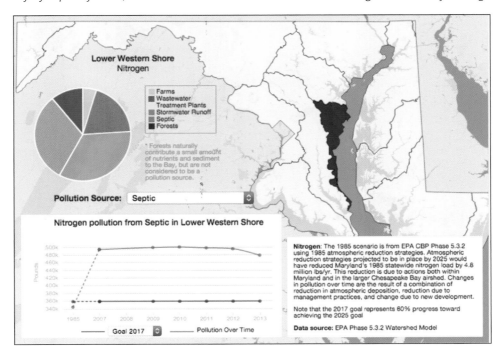

Where CompStat mapped crimes and deployed policing assets to reduce crime, BayStat mapped the sources of water pollution and deployed targeted actions to reduce it. Over time, the maps, the images, and the data became more timely, accurate, and clear.

The four main sources of Chesapeake Bay pollution are agricultural runoff, wastewater treatment plant discharge, septic system leachate, and stormwater runoff. The degree to which each of these sources of pollution affects any single rivershed or streamshed varies according the types of land uses or location of treatment plants within a rivershed.

For example, on the densely populated Western Shore of Maryland, stormwater runoff from hard-paved surfaces, wastewater treatment plants, and septic system leachate are the dominant sources of water pollution. On the Eastern Shore, agricultural runoff is the dominant source. And there are, of course, different actions that are needed to address each type of pollution.

For two of these sources—agricultural runoff and discharge from wastewater treatment plants—we had a good track record of taking actions on land that had been working for years. In these cases, the big challenge was to take our efforts to a more impactful scale.

A BayStat meeting in progress in the StateStat Room, with a number of different departments, July 26, 2011. Every month—without fail—for eight years, we convened the BayStat Cabinet to focus on driving the most effective leading actions to restore the health of the Chesapeake Bay. This was one of the largest collaborative Stats, with secretaries from the departments of natural resources, agriculture, corrections, transportation, the environment, and planning present and accountable at every meeting. Science was also represented at the table every month: Dr. Don Boesch from the University of Maryland Center for Environmental Science was present to call balls, strikes, and fouls when theories ran contrary to the science.

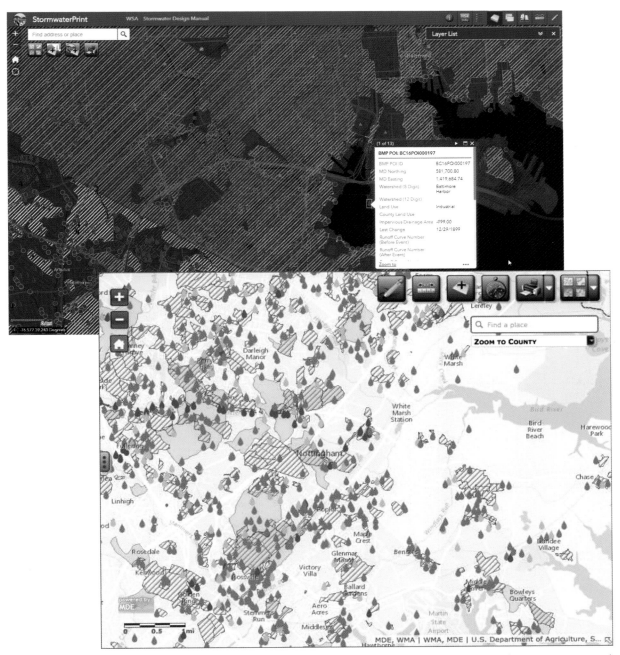

No entity was responsible for more blacktop asphalt surface than our own state Department of Transportation and Highways. StormwaterPrint is a web application developed by Maryland's Department of the Environment to map, track, and drive public and private stormwater remediation projects across our state. In this case, timely and accurate information shared by all included the locations of various stormwater remediation projects, as well as water quality data from monitoring sites.

For the other two main sources of water pollution—stormwater runoff and septic system leachate—positive action had never really been taken before. There were no rules to prevent and no frameworks with funding to remediate. Unchecked stormwater runoff had been particularly harmful to the health of rivers like the Patuxent that flow through densely populated areas, while the unchecked proliferation of aging septic systems had done tremendous damage to the health of rivers like the Severn, the Magothy, and the South River on the lower Western Shore of the Bay.

BayStat: The Method

Having established the four primary causes and implemented a common platform for action, we began to run plays. Around the BayStat map, we gathered the Bay Cabinet and other participants in a collaborative circle every month to measure performance and drive leading actions to our goal.

Seated around the BayStat table were the governor, director of StateStat, deputy chief of staff for Bay departments, chief innovation officer, budget director, and governor's legal counsel. First among cabinet equals was the secretary of Natural Resources. And on either side of him around the table were the secretaries for the Departments of the Environment, Agriculture, Transportation, Public Safety and Correctional Service, and the director of the Critical Areas Commission. Preparing the agenda and keeping follow-up notes was the dedicated staffer from StateStat.

Governing for Results: BayStat

We were not the first administration to pursue the goal of restoring the Chesapeake Bay, but no administration before made greater strides. We put in place a framework for success, including:

- The actions and funding mechanisms necessary to reduce all four major sources of pollution from land.
- BayStat, the performance measurement and management system that drove the restorative actions taken on land in ways that all citizens and stakeholders could see, monitor, and guarantee.
- The agreement by all six neighboring states and the EPA to take necessary, verifiable, and measurable actions on land, accounted for against two-year milestones, to restore the waters of their own state that flow, ultimately, into the Chesapeake Bay.
- Over the course of eight years, we accomplished measurable change—guided by science, and made open and transparent for all to see thanks to the relatively new application of GIS and the internet.
- The health of the Chesapeake Bay was improved for the first time in 350 years. Even greater improvements are possible in the future, if we continue to do what works on land to reduce pollution and improve the streams, rivers, and waters of the Chesapeake Bay.

Like CompStat or CitiStat, every meeting had an agenda, and notes were taken at every meeting on commitments made since the last meeting to solve problems, unravel bureaucratic tangles, or bring better tactics to bear on any given challenge.

We started, and we never stopped. We lifted up the leaders. And we led.

Unlike prior efforts, we set two-year milestones for achieving certain pollution reductions. We developed and publicly released a delivery plan with trajectories charting our course for the ensuing months and years that would take us to our goal with ever-greater reductions of nitrogen, phosphorus, and sediment.

We created a public-facing dashboard for BayStat broken down by each of the riversheds so that citizens over the internet could track progress, or the lack of it, with the same dashboard views of BayStat that their governor and cabinet used.

In meetings, we lifted up the leaders of those soil conservation districts that had been most successful at signing up farmers to implement best management practices or to enroll in the cover crop program—a program that paid farmers to plant winter crops after a season of corn or beans.

Chesapeake Bay Health Indicators

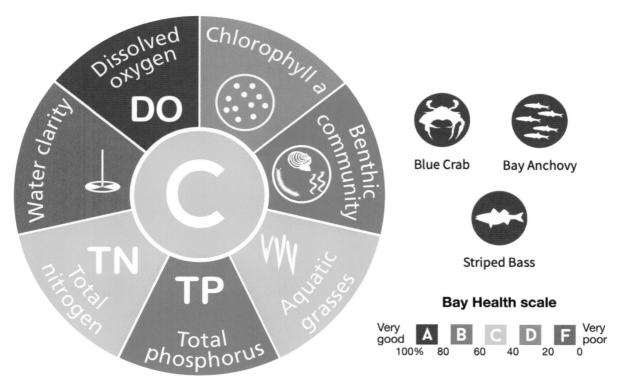

The Chesapeake Bay health report card compares seven indicators (dissolved oxygen, nitrogen, phosphorus, chlorophyll a, water clarity, aquatic grasses, and benthic community) to scientifically derived thresholds or goals. These indicators are combined into an Overall Health Index (on the left, an overall grade of "C" in 2017). On the right, blue crab, bay anchovy, and striped bass are also monitored but not included in the Overall Health Index score.

Bringing Back Native Oysters

Decades and decades of pollution, disease, and overharvesting led to the near extinction of the native Chesapeake Bay oyster. One of the bigger changes we made in the management of the Chesapeake fishery resources during my administration was a focus on actions to bring back this iconic and essential resource.

Any student of the Chesapeake Bay ecosystem learns quickly that the native oyster is not only a food resource, but the essential natural water-filtering species for the entire Bay.

For literally a hundred years, *Baltimore Sun* editorials had been urging a switch from a hunter-gatherer approach that extracts and depletes the oyster to an aquaculture approach where we "plant" more than we harvest.

In my administration, we finally made this change. And it worked. For the first time, we established tributary-wide oyster sanctuaries, which eventually covered about 25 percent of the Bay. And we also built a surface radar system throughout the Bay, called the Maritime Law Enforcement Information Network (MLEIN), to police these sanctuaries and protect them from poachers.

We turned the muddy bottom of Harris Creek, on the Eastern Shore of the Bay, into the largest oyster replenishment project undertaken anywhere in the United States. This project involved purchasing literally tons and tons of oyster shell from an ancient quarry in Florida and then transporting train loads of it to the Port of Baltimore. From there, the shell was taken by barges to Harris Creek, where it was spread across the large creek bottom to create a healthy substrate. Finally, millions of newly spawned oyster spat from the University of Maryland's Horn Point Lab Oyster Hatchery—a hatchery whose production capacity we greatly expanded—were planted to grow on this new healthy underwater habitat in Harris Creek.

The project was praised for its sound science and data-driven approach. Detailed bottom surveys and oyster population surveys were conducted to best target restoration efforts. And the level of cooperation among federal, state, nonprofit, university, and commercial partners was unprecedented.

Restoring the native oyster to filter the waters of the Bay and growing the strength of this fishery to support more jobs and more opportunity was met with near-universal acclaim by scientists, the environmental community, and the public. While many factors impact oyster population and harvests, the successes of this project prove that restoring oysters to fulfill their historical ecological role while rebuilding their economic value are mutually reinforcing goals.

With Governor Harry Hughes to my right at the state's Horn Point Lab Oyster Hatchery. In the late 1970s, Governor Hughes challenged all of us to "Save the Bay." In 2007, together, we expanded the capacity of this hatchery a hundredfold. By 2014, we had produced and planted a record 1.25 billion native baby oysters ("spat") in substrates of Maryland's Bay and had begun the largest replenishment and aquaculture effort in the nation.

The cover crop soaks up the remaining nitrogen and phosphorus fertilizer and holds the soil until the next planting season.

We used the meetings to break down and unpack the reasons for departmental resistance. We also used the meetings to communicate, collaborate, and cooperate about the leading actions that work to expand stream buffers, accelerate upgrades to wastewater treatment plants, or upgrade aging septic systems.

Science at the BayStat Table

There was one very important seat at the BayStat table, and I did not appoint its occupant. That seat at the table was reserved for *science*. Attending every BayStat meeting was the head of the University of Maryland Center for Environmental Science (UMCES), Don Boesch. UMCES was invited, but they didn't have to come. They had academic independence, so they had some insulation from political pressure and political expedience. They were free to tell us the truths we needed to hear.

The job of UMCES at the BayStat table was to call, as best they could, the balls and strikes based on the latest science and engineering. They answered questions such as these: What is an acre of cover crop worth in nitrogen uptake after a season of corn? What is the daily benefit of enhanced nitrogen

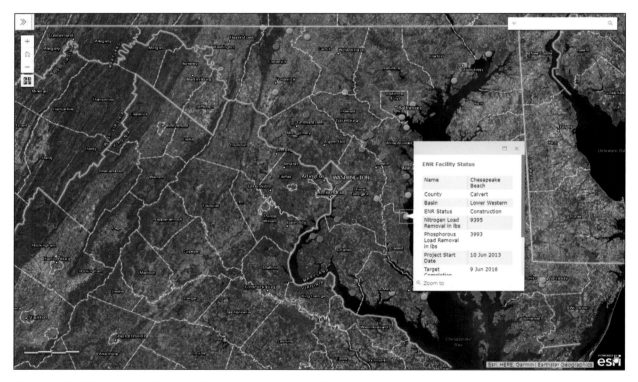

The so-called $5-a-month "flush tax" allowed us to modernize wastewater treatment plants across our state to greatly reduce the amount of waste discharged back into the watershed. After they were upgraded, plants with Enhanced Nutrient Removal (ENR) technologies reduced nitrogen, phosphorus, and other pollutants to record lows. This interactive map allowed citizens to track the status of ENR upgrades at wastewater treatment plants across the state. With everyone watching, these critically important capital projects were delivered on time and in time.

removal at a wastewater treatment plant compared to the existing technology being used? And how do we measure water health?

Many of these questions had been asked and answered already. But never before had those answers become part of the monthly deployment tactics, strategies, and actions with the governor and cabinet secretaries present and focused. Never before had science been in the center of a collaborative, iterative, and performance-managed process for Bay restoration. Never before had we really applied the math to solving the problem.

Sure, there had been plenty of reports over the years. Shelves were full of them. But BayStat applied science to the iterative, operational routines of a collaborative state government focused on the "wildly important goal" of restoring the health of the Chesapeake Bay.

UMCES never missed a meeting. And the scientists there had by this point developed a clear and compelling scoreboard for measuring the health of the Bay. They rolled it out the year before our administration was elected. Their map of the major tributaries of the Bay was clear: a green, yellow, orange, and red visualization of the health of the different riversheds, scientifically measured by objective criteria.

Checking out the shoreline critters with kids at Bishops Head, Dorchester County, just south of Blackwater Wildlife Refuge. The Bay Health Index is a combination of water quality indicators—dissolved oxygen, chlorophyll a, and water clarity—and biotic indicators—benthic life, phytoplankton, and aquatic grasses.

This Bay Health Index was composed of six measurable criteria. Water clarity, chlorophyll a, and dissolved oxygen combined for the Water Quality Index, while the measure of benthic community, aquatic grasses, and phytoplankton community combined into a Biotic Index. These scores were combined into one Bay Health Index for each major tributary or rivershed. And then the overall score for the entire Bay was derived from the sum and average of these rivershed scores.

In 2006, the Bay Report Card, as UMCES called it, very publicly gave the health of the Chesapeake a D+. On the report card map, her waters were a sea of mostly yellow, orange, and red—red being completely dead. There was no rivershed that was colored a healthy green.

We set out to change that.

Driving Leading Actions at BayStat Meetings

The Bay Report Card provided a clear, annual progress report on the *lagging indicators* of water health throughout the Bay. It became a definitive, objective, and scientific annual physical of the Bay and her tributaries. It was a valuable measure of the latest reality. But it was not a measure of the leading actions needed to improve that health.

The purpose of monthly BayStat meetings was not to stare at last year's health, but rather, to drive the leading actions that improve the health of the Bay this year. With the backup of our federal government and the Environmental Protection Agency (EPA), we put the bay on a diet—a maximum daily load of pollution calories, so to speak. These calories came not from Fritos or Doritos but from nitrogen, phosphorus, and sediment.

Our wildly important goal was clear—we needed to improve Bay health by reducing the flow of these pollutants into the tributaries of the Bay. Science and math told us the most impactful thirty-three leading actions we needed to focus on to reduce those flows and improve Bay health. The monthly BayStat meetings established a cadence of accountability for making sure those actions got done. And the BayStat website, with its dashboard of graphs and trajectories to two-year milestones, provided us a compelling scoreboard so everybody could see whether we were winning or not. It was a compelling scoreboard that every stakeholder and every citizen could see.

Driving these leading actions was the focus of every BayStat meeting. And a typical conversation in the course of the agenda might go something like this:

Question: "Mr. Secretary of Agriculture, are we on track for hitting this year's goal for cover crop enrollment?"

Answer: "At present, we are running short of goal and just about even with last year."

Question: "Why?"

Answer: "Not exactly sure. We are doing everything we did last year, but it's possible that we have kind of saturated the market. In other words, we think we have signed up every acre from just about every farmer that is willing to participate in the voluntary program."

Question: "Can you show me the number of active acres under cultivation by soil conservation district, and the percentage in each district that is enrolled in the cover crop program?"

Answer: "I can find out, but I don't have it today."

Question: "Thanks. Find out. We would all like to see it on a map next meeting."

At the next meeting...

Answer: "There would appear to be some wide variations in percentage of active acres from one soil conservation district to another."

Question: "Which district has enrolled the highest percentage, and who is their leader?"

The Four Disciplines of Execution

One framework I have found extremely useful in getting people's collective imagination to embrace the solving of big problems comes from a book called *The Four Disciplines of Execution* by Chris McChesney, Sean Covey, and Jim Huling. Their four disciplines of execution framework is all about focus and follow-through. And this requires both leadership discipline and organization discipline.

Discipline 3
KEEP A COMPELLING SCOREBOARD

Discipline 1
FOCUS ON THE WILDLY IMPORTANT

Discipline 2
ACT ON THE LEAD MEASURES

Discipline 4
CREATE A CADENCE OF ACCOUNTABILITY

In the context of actions to improve the health of the Chesapeake Bay, our **wildly important goal** was to reach the Healthier Bay Tipping Point by 2025— that point in time where the various rivers flowing into the bay would start getting healthier every year rather than sicker.

We **acted on the lead measures**—those thirty-three things we could do on land to reduce the nitrogen, phosphorus, and sedimentary pollution flowing into the bay.

We kept a **compelling scoreboard** that measured both the value of our actions—the pounds of pollution our actions reduced—and the lagging measures of bay health— objective scientific measures such as water clarity, dissolved oxygen, and phytoplankton life. And we made those real-time measures available in easy-to-read dashboards, maps, and graphs on the BayStat website.

Finally, with BayStat, we created a **cadence of accountability**—short monthly meetings of all the stakeholders, including the best scientific advisors from academia, to focus on the challenge before us and the actions—on the map—that would drive us to our wildly important goal.

In 2016, the waters of the Chesapeake Bay were the healthiest they have been since 1985. And we can keep making them healthier if we stay focused on the wildly important goal, if we act on lead measures, if we keep a compelling scoreboard, and if we maintain a cadence of accountability.

The difference between a dream and a goal is a deadline, and the difference between declaring a goal and achieving a goal is the four disciplines of execution.

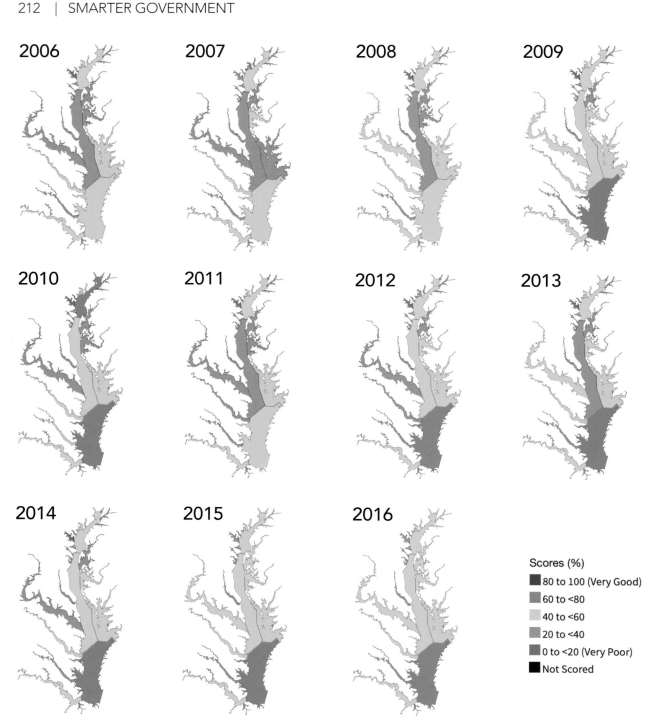

Chesapeake Bay health report cards from 2006 to 2016 (above) and 2017 (right) show the longest sustained improvement in Bay health in 350 years. The red-to-green color scheme made the results of our shared work understandable by all. Red is dead. Orange is dying. Yellow is sick. Green is healthy and living. The trend is now toward life. For more information, visit chesapeakebay.ecoreportcard.org.

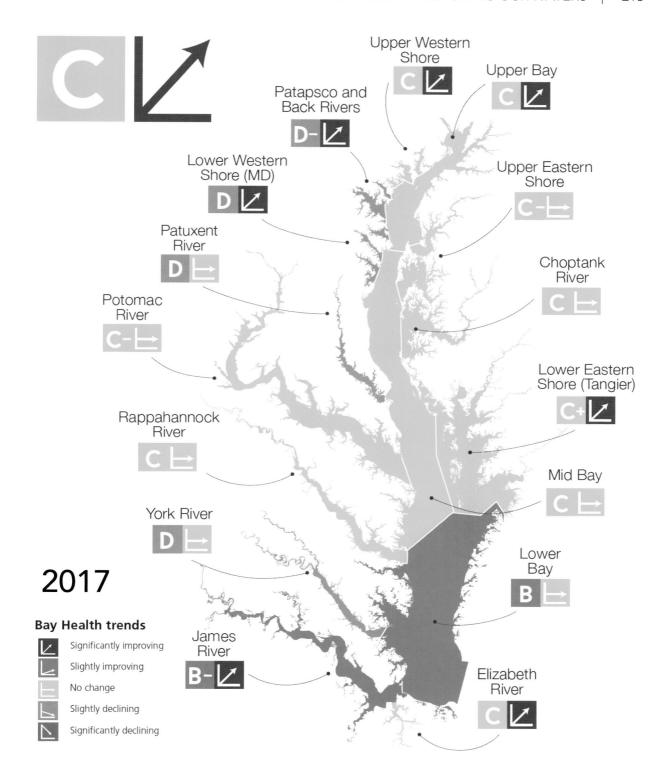

C

Upper Western Shore
C

Upper Bay
C

Patapsco and Back Rivers
D-

Upper Eastern Shore
C-

Lower Western Shore (MD)
D

Choptank River
C

Patuxent River
D

Potomac River
C-

Lower Eastern Shore (Tangier)
C+

Rappahannock River
C

Mid Bay
C

York River
D

Lower Bay
B

2017

Bay Health trends

Significantly improving

Slightly improving

No change

Slightly declining

Significantly declining

James River
B-

Elizabeth River
C

Answer: "Betty Sue in Southern Maryland."

Question: "What is she doing that the other districts are not?"

Answer: "Governor, we are asking the same question. It seems she makes a lot of follow-up calls..."

Question: "Good. Sounds like everybody needs to try making more follow-up calls. Let me know what else she does to be so successful. Share it with the others. And send me her email after the meeting. I want to send her a thank you note. And I'll cc you, so you can forward it, please, to everybody else in Soil Conservation, okay?"

Answer: "Yes, sir."

And so it would go.

We never stopped asking those types of questions about the effectiveness of our leading actions. Actions that enrolled more acres in cover crops. Actions that expanded stream buffers. Actions that lead to the installation of better management practices on farms. Actions that replaced hard surfaces with permeable surfaces in urban and suburban settings. Actions that upgraded aging septic systems and wastewater treatment plants.

The conversations were not always smooth and enlightening. Sometimes the conversations were painfully frustrating. But over the course of many months and years, the repeatable routine of meetings and the focus on leading actions allowed us to achieve our two-year milestones as well as our longer-term goal—a Bay whose rivers started getting a littler healthier every year instead of a little sicker.

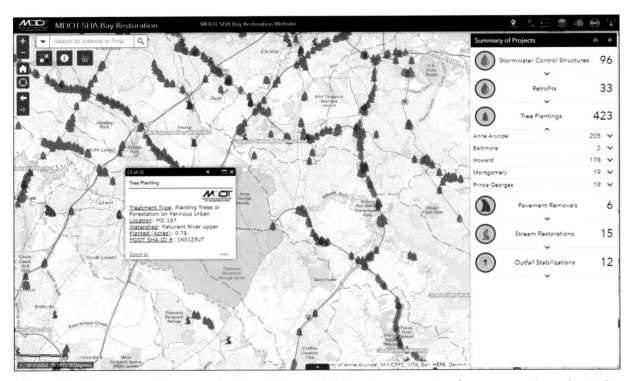

The Maryland Department of Transportation State Highway Administration uses a series of proven strategies and controls to reduce nutrients and sediments that reach local waters and, ultimately, the bay. This dashboard shows the location and status of projects—including stormwater control structures, retrofits, tree plantings, pavement removals, stream restorations, and outfall stabilizations—being implemented to meet pollution reduction goals to help restore the Bay.

Leading Actions on Land to Improve the Waters' Health

To achieve our goals of nitrogen reduction, we focused on the thirty-three leading actions that we needed to take to a more impactful scale on land to improve the health of our waters. We measured, modeled, and mapped the amount of nitrogen each individual action—taken to targeted scale—would remove from the watershed.

In a break with traditional political norms, we didn't set some far-off deadline that wouldn't be measured until my term of office ended. Instead, we set two-year milestones by which every voter in the state could assess whether we were making the progress we promised. Good leaders make themselves vulnerable. It's the only way to get difficult and important things done.

These leading actions spanned all types of land uses in our state—urban, suburban, and rural. Everybody was in, everybody was at the table, and everybody was needed.

The actions broke down into three large categories—improving farming practices, reducing pollution from developed lands, and restoring natural filters. With its real-time dashboard, the BayStat website measured the actual trajectory of our progress on any given action against our planned trajectory for reaching our goal. These results could be viewed in either a statewide format or by individual riversheds, that is, the Potomac riversheds, the Patuxent, the Choptank, and so on. This openness and transparency allowed all stakeholders and every citizen to measure the progress of our leading actions in real-time, and to compare the trajectories of progress from one rivershed to another.

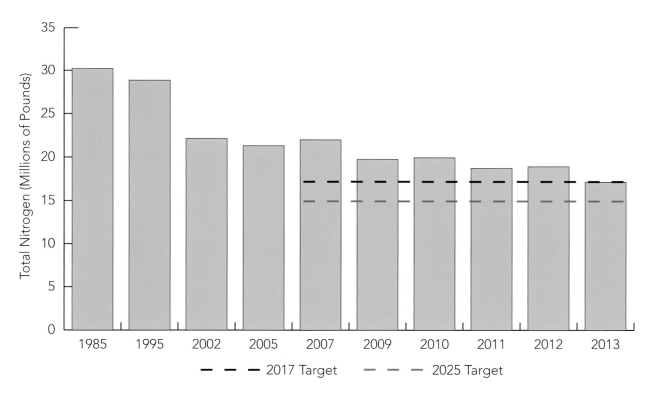

After meeting our 2017 target for reducing agricultural nitrogen load four years ahead of schedule, we set a new target for even greater reductions by 2025.

There is a natural beauty and mystery to the winding waters of the Chesapeake Bay, but the actions we can take on land to make those waters healthier rather than sicker is really no mystery at all. The data, the map, and the method of relentless follow-up are key to forging and holding a more life-giving consensus to make it happen.

The charts showed the overall goals for total nitrogen reduction for the two-year milestone ending in 2012. The BayStat dashboard showed the trendline and trajectories for each individual action. The maps showed us where we were taking these actions in any given year, and where the opportunities were for taking further actions to reduce nitrogen flows.

State and National Action to Help the Bay

Maryland was the first state in the Chesapeake Bay watershed to apply this new way of governing to the great work of restoring the waters of the Chesapeake Bay. But we would not be the last. We showed that it was possible. Setting up a performance management system was not technically difficult or expensive. It was common sense and it was a much more effective way of getting things done with openness, transparency, and accountability.

A year after Maryland's launch of BayStat, our neighbors, the Commonwealth of Virginia and the District of Columbia, stood up their own BayStat systems to drive leading actions within their borders of the Bay watershed. The fact that Maryland, DC, and Virginia were now able to measure and show progress soon had the other states taking notice.

And when President Barack Obama was elected, the waters of the Chesapeake Bay were the subject of his first executive order for the cleanup of a major body of water in the United States. He committed the federal government not only to greater action, but to greater involvement. The occasion gave us the ability to persuade all six states in the watershed to commit to two-year milestones and measurable actions. And to track and monitor commitments throughout the six-state watershed, the EPA soon stood up its own version of BayStat, dubbed "ChesapeakeStat."

In a voluntary agreement, the governors of all six states bordering the Bay asked the EPA to back up our own commitments with enforcement actions—if necessary—to make sure we hit our goals of reducing water pollution from our various states. A total maximum daily load of nitrogen, phosphorus, and sediment was set for the entire watershed. And together, all of us agreed to start and not stop.

Reaching the Tipping Point

It was an unprecedented level of commitment to a new way of governing, measuring, and getting things done for the improvement of a major body of water.

There were stories within the larger story—native oyster restoration, saving the blue crab, the healing work of Maryland inmates on the Bay in our new restorative justice efforts, and the way Pennsylvania localized the effort to individual streams throughout the vast rolling mountains and countryside of their state.

With every state now taking measurable action, it wasn't long before we saw measurable improvements—not just in the scale of the actions we were all taking, but in the lagging indicators of Bay health as measured by the Bay Report Card.

Our goal had been to reach the Healthier Bay Tipping Point by 2025. Rainfall levels and drought were things beyond our control, and their occurrence meant that progress rarely happened in a straight line or without a year or two of setback. But our actions we could control. We could map them, and we could drive them to a more impactful scale. And so we did.

The political challenges were many. We increased the monthly flush fee (assessed on any occupied home with one or more bathrooms), we halted the proliferation of large housing developments served by outdated septic systems, and we put in place a fee system to encourage the remediation of harmful stormwater runoff in ten of Maryland's most densely populated counties. Each of these measures was controversial. None of them was accomplished easily. I still bear the political scars in the eyes of some of my neighbors who were convinced by political opponents that I was so tax happy I wanted to tax even the rain. In a strange irony, the citizens most easily worked up against the so-attacked "rain tax" lived in more rural counties where no stormwater tax or fee applied.

This is what it looked like, over the progress of just one decade as the map moved from a declining orange-to-red trend, to an improving yellow-to-green trend. The Bay Report Card for 2016, for the first time, showed steady or improving trends in every single rivershed of the Chesapeake Bay.

Most Bay scientists believe we reached that tipping point in 2016—nine years ahead of our 2025 Goal.

Leadership. A new way of governing. Getting things done. We made more progress more quickly than many of us thought possible. And Mother Nature was even more resilient than we thought she could be.

The great work continues. In fact, it's never really going to be done. But after 350 years of decline, at least we are now moving in the right direction.

Certified Cover Crops (Acres Planted)	Stormwater Retrofits (Pounds of Nitrogen Removed)	Natural Filters on Private Lands (Acres)
FY 2014	FY 2013	FY 2013
410,530 Acres	221,576 Pounds	110,035 Acres
FY 2006	FY 2006	FY 2006
124,800 Acres	74,706 Pounds	79,519 Acres
229.0% INCREASE ▲	196.6% INCREASE ▲	38.4% INCREASE ▲

A StateStat dashboard showing progress towards reaching the state strategic goal of reaching the Healthier Bay Tipping Point by 2025—that point at which our better actions on land would join forces with the restorative powers of nature to make all of our riversheds a little healthier every year instead of a little sicker every year (as of July 2, 2014).

Learn & Explore

Chesapeake Bay DataHub
The Chesapeake Bay DataHub offers access to a multitude of monitoring and modeling data, past and present.

Maryland BayStat
The primary goal of this website is to engage Maryland's citizens in restoring the Chesapeake Bay by providing a transparent presentation of our best, scientifically based assessment of the health of Chesapeake Bay and its tributaries, the sources of the problems, and the effectiveness of the wide variety of programs designed to address the problems and restore the Bay.

Water Quality Status Dashboard
This interactive map-based dashboard provides a visual status of water quality conditions in Maryland.

For links to these and other examples, exercises, and resources, visit SmarterGovernment.com and click chapter 10.

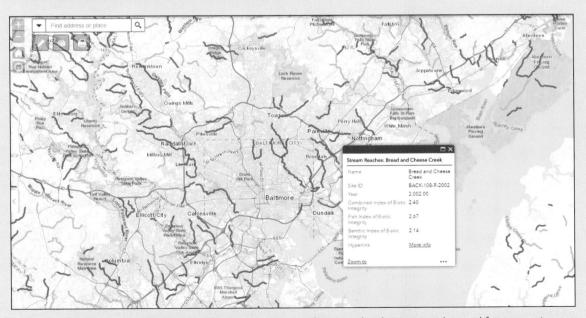

Stream Health is an interactive, crowd-sourced map providing Maryland citizens with a tool for measuring, assessing, and tracking the health of the smaller neighborhood streams throughout our state. The map also shows citizens where stream buffers are present or not, and where impervious blacktop surface is concentrated in any streamshed—two important determinants of stream health.

11

Preserving Our Land

We are surface dwellers. Our lives and our well-being depend on the health of the land, air, and waters of this earth. And yet, throughout recorded history, we have treated the land and the landscape as a commodity to be used rather than a living system upon which our children's lives depend.

A new human ethic is dawning. It is the ethic of the land—or perhaps the ethic of the earth. And new technologies—like GIS—give us the ability to preserve it, to restore it, and to protect it.

Natural Lands, Open Space, and Our Quality of Life

We are not producing new land. Natural, recreational, and ecological land resources are a vital and utterly indispensable requirement for healthy, vibrant, livable communities at any scale. Protecting, preserving, and restoring our natural land resources—our green infrastructure—is not merely a discipline in land management; it is essential to reversing climate change and protecting human life on this planet.

To do this, we need to give governments and citizens easy access to a comprehensive picture. A picture that maps clearly for all to see our plan to protect green infrastructure—our GreenPrint—and whether the actions we take are protecting and restoring it over time.

Mapping Valuable Land: Maryland's GreenPrint

In my home state of Maryland, the words "land use" are fighting words on the battlefield of state and local politics. Few things rile up the crowd and spoil the fun at any annual meeting of the Maryland Association of Counties quite so much as bringing up land use issues. Especially if a governor—any governor—tries to tell county officials where they should and shouldn't allow people to build. For many elected officials at the county level, these land-use and zoning decisions are jealously guarded prerogatives—the freedom to "do what we want" with our land.

But here is the problem. More land has been consumed by development in the State of Maryland over the last thirty-five years than we consumed in our first 350 years as a state. Calls for "smart growth," limits on sprawl, and greater sustainability are rising. But the growing awareness of this new ethic, this land ethic, is still relatively new—especially for the western mind.

Governors who try to dictate to their counties where growth should and shouldn't happen really have their work cut out for them. It's a huge political lift. Better to lead with protecting water and open space than with dictating growth zones. Better to protect food supplies and family farms than proscribe growth limits.

Let me show you what I mean. Imagine, if you will, a blank page. This page represents the land of a state. It could be any state prior to colonization. With European settlement, western expansion, Manifest Destiny, and population growth, the page started to fill up with development.

It wasn't as if there were no humans around—the Piscataway and Conoy and other native peoples had lived in the area we now call Maryland for centuries. But this new European culture had a different concept of the land and how to use it up.

Forests were cut down. Towns sprang up and became cities. Cities became metropolitan areas. The patterns of land use and development began at seaside ports, and then followed rivers, and then rail lines, roads, and sewer lines. County seats dotting the countryside expanded and grew into a random, patchworked hodge-podge that failed to account for environmental impact and generational sustainability.

Suburbs expanded where once there were farmlands, woodlands, and wetlands. Swamps were filled in for development. Farms became housing subdivisions. Soon you couldn't move away from one big city without hitting the suburbs of the next big city.

Growth for the sake of growth seemed—for a time—a pretty virtuous cycle.

Then a time came—this time. We suddenly came to understand that land is a finite resource, and the blank spaces on the map just might be the most valuable. Now we realized we need the land more than the land needs us.

In our efforts to advance the smart growth agenda begun in Maryland under my predecessor,

Governor Parris Glendening, we did not lead with restrictions on growth. Instead, we led with protections of green infrastructure. We set out to uncork the previously closed datasets (most of them federal), with Esri's help and collaboration, that would allow us to measure, rank, and map the most ecologically valuable parcels of land left in our state.

The result was Maryland GreenPrint—a comprehensive map of green infrastructure ranking the most ecologically valuable parcels of land in the state in a way that all could explore and see. This map became the left margin, if you will, of our previously blank page for future development.

By mapping Maryland's GreenPrint, we essentially moved the left margin farther toward the center of the page. We showed every county, right down to the individual neighborhood and parcel level, where their GreenPrint was. We showed every citizen which parcels were already protected, and which parcels still needed to be protected to buffer streams, connect habitat corridors, and preserve the ecological well-being of this place we call Maryland for future generations.

Everyone applauded—even the local county officials.

Better management practices implemented by hundreds of farm operations throughout Maryland have led to healthier soils and healthier waters as sustainability and profitability for farmers join forces for a cleaner Bay.

Preserving Farmland: Maryland's AgPrint

After we had a GreenPrint, we then began to move the margin of the right side of the page toward the center. It was an exercise similar to GreenPrint but with a slightly different purpose. We measured and mapped the most valuable remaining agricultural lands left in our densely populated state.

This became Maryland's AgPrint—the family farms and contiguous agricultural economies we needed to protect to grow the food we needed to survive.

We announced this map at the Maryland State Agriculture Dinner. Like GreenPrint before it, AgPrint was a big hit. We could all applaud preserving farming in Maryland. On an intuitive level, we all knew this was in the best interest of the common good we share as a people. And now everyone could see for themselves, online, which parcels had been preserved and which ones had not yet been preserved, but should be.

By this point, as you might imagine, the space remaining down the center of the page had been greatly narrowed, and so, too, had the political battlefield. Prescribing where future growth should take place became a lot easier when we could all see so clearly where it should not.

AgPrint uses GIS to analyze Maryland's Resource Lands, which consist of natural resource lands and other resource conservation lands outside targeted development areas. When every citizen can see where agricultural economies exist, local land-use decisions become better informed by the common goal we share of protecting farmlands and farm livelihoods for food production.

An Idea for Buying Land: Program Open Space

Back in the early 1970s a visionary state senator named James Clark, from then-rural Howard County, saw how quickly land in Maryland was being consumed by suburban sprawl and development. His family had been farming for generations in Howard County, which was right between the expanding Baltimore suburbs to the north and the expanding Washington, DC, suburbs to the south, located in the middle of what is now the most densely populated corridor on the East Coast.

Senator Clark saw that something had to be done to preserve the natural and agricultural lands of Maryland before these lands were no more. His solution was Program Open Space—an initiative that would help preserve Maryland's natural and agricultural lands—our green infrastructure—using all the revenues from state transfer taxes. It was a simple and elegant idea. When there was any sale of land anywhere in the state, the state revenues generated by that sale went to a fund to purchase other lands in Maryland for preservation.

It was one of the first and most comprehensive land preservation statutes passed by any state in the nation. And it was at this point that Maryland became client number eight for a then-new, little GIS company called Environmental Systems Research Institute (now known as Esri). Public administrators in Maryland, realizing they were going to use public dollars for the purchase of private lands, needed a way to keep track of the "where and how much" of all those transactions. GIS provided the solution.

GIS also provided the solution for tracking lands protected through other initiatives, like the Maryland Agricultural Land Preservation Program and the Maryland Environmental Trust—programs that primarily exchange property tax abatements for the transfer of private development rights.

It would be many years before the same map that tracked parcel, place, and purchase price would begin to be used to track an ever-expanding wealth of information: data about the ecological value of any given parcel—its biodiversity, its forest cover, its natural ability to provide a protective stream buffer or a habitat corridor for birds and other wildlife. Data that showed the value of particular parcels for the health of the waters of the Chesapeake Bay. These were some of the things Maryland GreenPrint made possible.

Today, every time staff from the Maryland Department of Natural Resources recommends the purchase of private land through Program Open Space to the Maryland Board of Public Works (made up of the governor, the comptroller, and the treasurer), the recommendation is accompanied by key information everyone can see. This information includes not only the maps showing where and how a particular parcel is connected to the larger Maryland GreenPrint, but also an objective scoring criterion based on the parcel's ecological value.

The scale ranges from as low as zero to a high of 150. A low score on habitat, for example, might be offset by a high score for public access to water. Most purchases fall within a range of 70 to 120. Decisions about which lands to purchase and how much to pay for them can now be based primarily on empirical scientific evidence about the parcel—on its connection and value to the whole—and not simply on expediency or how politically connected the seller might be. And by displaying all this online, on maps that every citizen can explore and query, the public and their journalists can see whether their public dollars are truly going for a public purpose.

One of the more memorable purchases I approved as governor was accompanied by a Papal Imprimatur. It was the purchase of seven hundred acres of land at three different sites near the St. Marys and Bohemia Rivers. The properties had only one recorded owner in the 350-year title search: "The Corporation of Christian Gentlemen," the Jesuits who accompanied the first English Catholic colonists to Maryland in 1634.

Tackling the Need for Park Equity

Although it's been more than fifty years since the Watts riots near Los Angeles took place, the aftermath of the unrest can still be seen today. In places, it appears as if entire neighborhoods have been dismissed due to the violent six-day episode in the distant past. Lack of investments in the community has had a significant impact on the quality of life and the health of Watts residents.

Ronald "Kartoon" Antwine, a reformed gang member and lifelong Watts resident, wonders: "How come when I leave my neighborhood, I see clean streets and greenery, but when I come back here, I'm looking at trash and weeds?"

One small strip of land at the corner of Monitor Avenue and 114th Street symbolized the painful effects of such urban blight. Weeds grew to seven-foot heights, and the empty space was prone to illegal dumping of unwanted stuff. The Trust for Public Land, a nonprofit organization that partners with cities nationwide to transform vacant land

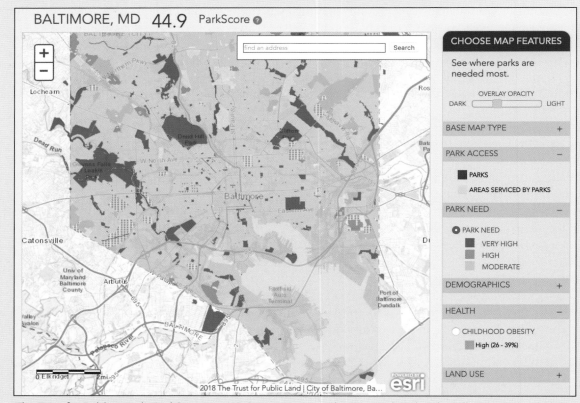

The Trust for Public Land's ParkScore uses GIS to examine and promote the idea that all citizens should have a park within a ten-minute walk.

into parks, worked with the local community and the City of Los Angeles to acquire the property, identify community priorities, and turn the vacant lot into a park.

While Watts is one of the most park-poor areas in Los Angeles County, it's not alone in needing more parks and green spaces. Like most metro areas, cities across the county have struggled to maintain and improve citizen access to parks. The county is the most populous in the nation and one of the largest, with ten million residents across eighty-eight cities and an area of more than 4,751 square miles. There are 3,024 parks that cover nine hundred thousand acres; however, most of the acreage is currently in mountain parks, and urban centers are left lacking.

The Trust for Public Land has been working in Los Angeles for decades. It was the first place the organization chose to apply GIS software to help map priority places for parks. "I started in 2001 as the first GIS person hired," says Breece Robertson, now the organization's vice president and cheif research and innovation officer. "My first job was to figure out how to measure equitable access to parks in Los Angeles.

"We started with very rudimentary analysis—half-mile buffers around parks, population density, and demographic data," says Robertson. "We now have sophisticated models to determine who has access to parks and who doesn't, leverging 10-minute walkable service areas and rich socio-economic data. We use GIS to overlay demographic, health, and climate justice information to determine where to straegically focus our attention.

"We're focused on the communities that need the parks the most," says Robertson. "If there were two options on the table to work in a high-income area versus a low-income area, we would choose the low-income area where we'll have the most impact."

Outside of its project-based work, the Trust for Public Land has a broader effort to spread awareness of the value of parks. The organization's ParkScore® index, now in its eighth year, can be used to assess and compare the quality of park systems in the hundred largest US cities. The ParkScore index is calculated using advanced analytical capabilities of GIS plus the City Park Facts research data to examine and promote the idea that all citizens should have a high-quality, safe park within a ten-minute walk.

"We're using GIS to develop 'What Would It Take' tools," says Robertson. "We can tell cities that if you're at 35 percent served and want to get to 50 percent served, here's what it will take to get you there. It will show how many parks need to be added, and where they should go plus all the benefits those parks will provide to the community. The Trust for Public Land helped San Francisco and Boston achieve a 100 percent served status and many other cities are closing in on that goal. Parks are essential for healthy connected communities and we are elevating the importance of parks by leveraging data, research and science."

Whatever else might be said of the cruel twists of history that followed, it is our remarkable good fortune that this small piece of our state's green infrastructure was preserved intact for more than three hundred years—and now, thanks to the people of Maryland, those properties will be preserved into the future.

Zooming In on Maryland's GreenPrint

Development has consumed more land in Maryland over the last thirty-five years than it did in our first 350. The natural areas that remain comprise Maryland's green infrastructure.

Our purpose in creating Maryland's GreenPrint—our first-in-the-nation ecological ranking of every parcel in the State of Maryland—was to do a better job protecting the natural lungs, kidneys, and liver that the ecological body of our state needs to be healthy enough to maintain the health of us all.

Maryland's GreenPrint highlights and ranks the state's most essential ecological parcels. Using varying shades of green, this online map shows what we've protected so far, and it identifies what else we need to protect through land conservation, preservation, or sound zoning and land use practices.

On the GreenPrint map, the areas in green identify the most essential ecological parcels left in our state. The dark green represents what we had protected through 2015. The light green represents what we still needed to protect through land conservation, preservation, or sound zoning and land use.

A citizen who clicked on Charles County would see the percentage of the county that's already been protected, the percentage of Program Open Space dollars that have been spent on purchases within the GreenPrint, and which purchases are outside of it.

Zooming in on the map, red stars hover above properties Maryland purchased with Program Open Space dollars within a given calendar year, and purple stars hover above land parcels purchased in another calendar year.

By zooming in further, we reach a level of detail where every citizen can see where Maryland's green infrastructure is in relation to their own neighborhood or their own home. The online map shows people how, why, and where we are protecting the GreenPrint in places closest and best known to them. And we have the option of viewing aerial photography as we zoom in, as well.

If we click on an individual parcel, we can see detailed information on things like the proximity of the parcel to other forest land or to stream buffers of a watershed, the date our state purchased the land, and how much state taxpayers paid to acquire and preserve it.

Maryland's GreenPrint became the foundational and guiding layer of the map of our state. It was not merely an example of how we used GIS technology to guide our management of natural resources; it became the foundational basemap for restoring the health of our land, air, and water for the greater well-being of all our people. It became the first map to consider when planning future development.

These strategies will work for any government anywhere. And, in fact, they already do—as more and more governments the world over harness the power of GIS technology to remake and restore their own piece of the surface of our planet.

Green Infrastructure in the United States

Whether we are talking about cities, counties, states, or countries, green infrastructure protection is fundamentally a spatial awareness problem. We cannot protect what we cannot see. But when we see it, we can not only protect it, we can restore it. We can plan for it. And we can plan around it.

Data about the natural world and our built environments—from many sources, in a variety of formats, and at a range of scales—can now be combined and modeled to uncover patterns, to see relationships,

Maryland's GreenPrint is a comprehensive map of green infrastructure in the state, ranking the most ecologically valuable parcels of land in a way that all can see, explore, and protect for future generations. "Show me my GreenPrint."

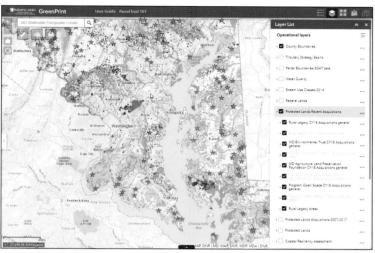

At the county level, we can see details about land that's already been protected, as well as lands targeted for purchase. Red stars identify properties purchased for protection within the current calendar year.

Zooming in further, every citizen can see where Maryland's green infrastructure is in relation to their own neighborhood or their own home. The online map shows people how, why, and where we are protecting land in places closest and best known to them.

Restoring a Habitat to Its Highest Value

In a corner of northern Minnesota, about halfway between Duluth and the Boundary Waters Canoe Area Wilderness, sits one of the most treasured and famous birdwatching habitats in the world—the Sax-Zim Bog. Highlighted in the book and movie *The Big Year* (starring Steve Martin, Owen Wilson, and Jack Black as jet-setting birders), the Sax-Zim Bog is one of the only places in the Lower 48 of the United States where one can see arctic and boreal bird species, including great gray, snowy, and boreal owls, as they are pushed out of Canada in the coldest months of winter.

Until just a couple of years ago, this vast habitat was both unprotected and highly degraded from almost a century of attempts to convert it into farmland and other more "useful" conditions.

Little did those bog-drainers of the 1900s understand that the highest and best use of these lands are for the value of the natural ecological services they provide—ecological services for the health of our people, of our waters, of our land, and of our air.

In 2016 and 2017, using entirely local expertise and labor, Ecosystem Investment Partners (EIP) invested millions in private capital to restore the bog by filling in the eighty miles of ditches, eradicating nonnative plant species, restoring the natural hydrology of the bog, and donating conservation easements over the land to the State of Minnesota's Department of Natural Resources. For almost three years, EIP employed thirteen full-time local laborers (many of whom were experienced, but out of work, heavy equipment operators who used to work in the shuttered iron mines just to the north) all at or above local union labor pay. EIP also paid a local environmental engineering firm for more than ten thousand hours (the equivalent to five full-time jobs) to design the restoration, determine how to restore the bog's hydrology, and measure the restoration success on the ground. As a private landowner, EIP pays property taxes to the county, and even agreed to establish and fund a payment in lieu of taxes (PILT) fund held by the local community foundation to ensure the project will always provide a sustainable revenue source for the local community.

Today, the Sax-Zim Bog has found new life—restored, protected, and even more cherished by the birds and people as an example of how smart capital, smart government policy, and dedicated conservationists can craft win-win-win solutions, all for the improved health of our land, air, and water.

—Nick Dilks, Ecosystem Investment Partners, LLC

Normalized Difference Vegetation Index (NDVI) aerial imagery for a portion of the Chesapeake Bay, made available by the USDA Farm Services Agency. The National Agriculture Imagery Program (NAIP) acquires this imagery during the agricultural growing seasons to assess vegetation health and makes it available to governmental agencies and to the public.

and to create a more holistic picture that everyone can see.

Currently, most local government entities that regulate the use of the land in the United States do so in isolation from one another, and often without tools to effectively collaborate across borders, assess existing conditions, or model the effects of proposed policy changes. But this is changing quickly.

GIS now makes better connected planning possible—and it makes it possible with better maps and better measures for public accountability than ever before. GIS provides the modeling views needed to integrate, analyze, and visualize a more sustainable vision for a place—whether that place is a nation, a state, a county, or a city. And it is this shared vision of the emerging truth that makes stakeholder communication more meaningful and collaboration more effective.

With Esri's help in opening previously closed and separate datasets, Maryland became the first state in the United States to create a statewide green infrastructure map—a map ranking every parcel of land in the state by its ecological value. And despite warnings that land speculators might use the open data for nefarious purposes, we intentionally published the map online so every citizen could see it—from a state, county, neighborhood, or household view.

Now, Esri has made the same green infrastructure map available for the entire United States—connecting landscapes, people, and communities across the country. Any state or county now has a running start. Every place can now create their own GreenPrint simply by assigning values or "weights" to parcels according to the ecological value, land use value, or other public policy goals, like agricultural land preservation.

Esri compiled, and published online, our nation's first green infrastructure map in 2016—a unique resource for green infrastructure planning at any level. The map identifies natural areas, or cores, across the United States larger than one hundred acres, regardless of ownership or legal preservation status.

The map shows areas of ecological, cultural, and scenic importance. People can change the size of these cores or rank them based on local conditions: hydrology, species, landform, elevation, flood plains, and other ecosystem data. Local and regional data can also be combined with the national map to evaluate the current state of an area and its probable future, given specific planning policies.

Green infrastructure provides economic, social, and ecological benefits. Space for food, space for clean water, space for quality of life. Knowing and seeing where these lands are on a green infrastructure map allows people to determine what actions are required to preserve these valuable landscapes.

The great democratic hope is that if everyone can see where our green infrastructure is, everyone can act to protect and restore it.

All Together Now: Protecting the Planet's Green Infrastructure

My friend Jack Dangermond—the founder of Esri and a landscape architect by his first trade—says humanity has reached a point in our journey with this planet where we must all learn to become "master gardeners."

We must realize that our lives depend on the other systems of this earth—especially the living systems, the plants, and the soils that comprise our green infrastructure. And we must take more life-giving actions in light of that growing understanding of our interdependence. Land is not only a finite resource that we hold in stewardship for our children and future generations; it is a living system upon whose health our own lives depend.

There is a Native American proverb that holds, "Whatever befalls the earth befalls the sons and daughters of the earth."

In other words: We are all in this together.

Understanding Green Infrastructure

Green infrastructure is a strategically planned and managed network of open space, watersheds, wildlife habitats, parks, and other natural and semi-natural areas that sustain life and enrich the livability of communities. Encompassing more than just roads, bridges, and rail lines that run across the land; green infrastructure is the connected natural lands themselves—the rivers, lakes, forests, marshlands, and grasslands. It stands in stark contrast to "gray infrastructure," which denotes highways, buildings, and other man-made features of the modern landscape.

A green infrastructure strategy identifies and connects our most valuable landscapes before new development begins. Why?

Because green infrastructure protects the health and diversity of wildlife and maintains natural systems that deliver critical, life-sustaining services.

Green infrastructure benefits property values, lowers health-care costs, and helps communities make smarter investments in traditional gray infrastructure.

Green infrastructure enhances quality of life by ensuring that people connect with nature, have access to clean air and water, and live healthier and happier lives. Balancing green and gray infrastructure is a recipe for building sustainable communities.

With a green infrastructure strategy like GreenPrint, communities can work to preserve and connect open spaces, watersheds, wildlife habitats, parks, and other critical landscapes for the benefit of people. A suite of green infrastructure planning tools is also available for use with the data behind the national green infrastructure map. Using these tools, planners, elected officials, researchers, businesses, and the public can explore the status of green infrastructure in a community now and extrapolate the effects of current planning on the landscape.

Applying a green infrastructure approach to ongoing efforts of sustainable land management encompasses these four steps:

1. Critically evaluate community policies on land development.
2. Benchmark the area against other communities.
3. Include a green infrastructure element in the community's comprehensive land-use plan.
4. Use GIS analysis and tools to map and measure progress over time as it relates to the preservation and restoration of a community's green infrastructure.

A Green Infrastructure Plan for Valencia, Spain

Geodesign couples geography with design to support designing with nature in mind and promote a healthy green infrastructure. One geodesigner making strides in the right direction is the Spanish landscape architect and urban planner Arancha Muñoz-Criado.

Muñoz-Criado has devoted much of her career to introducing land conservation and green infrastructure into the planning process in the Valencia region of Spain, where she grew up with a love for the "beautiful area of Spain—with its pristine beaches, mountains [overlooking] the seas, and wonderful terraces full of almond trees and vineyards."

But in the 1960s and 1970s, tourists from other parts of Europe discovered the fishing villages and beaches of Valencia, Spain. Soon, a crop of summer houses replaced many of the almond trees. "Suddenly, [the villages] grew very, very rapidly,... but development was allowed anywhere, and that was a total disaster," Muñoz-Criado said.

"I thought there was another way of growing—we could grow but grow well, preserving the character and preserving the landscape of the place," she said.

After earning a degree in architecture in Spain, Muñoz-Criado was accepted at Harvard University, where she earned her Master of Landscape Architecture (MLA) degree.

While in Boston with friends, Muñoz-Criado visited the Emerald Necklace, a seven-mile-long system of parks and waterways designed in the 1870s by American landscape architect Frederick Law Olmsted. "Every time I went there, I said, 'What a simple idea,'" Muñoz-Criado recalled. "Find out which places that you [want] to preserve before growing, and then develop around these places." She later brought those ideas home to Spain but realized that, to achieve policy changes, she would have to take a role in government to help enact them.

Moving up the ranks to eventually become regional secretary of urban planning, landscape, and environment, Muñoz-Criado eventually helped put new green infrastructure requirements into place—balancing social well-being, economic development, and environmental needs.

Today, the 550 municipalities of Valencia, Spain, now use geodesign to incorporate green infrastructure into the planning process. Ecological, cultural, agricultural, and flood areas are shown on the map. "Municipal planners, investors, and [other stakeholders]

Landscape architect Arancha Muñoz-Criado answers questions at the 2016 Geodesign Summit about the success of green infrastructue adoption for the entire region of Valencia, Spain.

know that in the green areas, they have some environmental restrictions," she said. "And they just have to click on the GIS map to know where they are."

Today, a plan is in place to create a green infrastructure network in the Valencia region that promotes clean air and biodiversity. Rules are set at a regional scale to protect forests, wetlands, and agricultural areas. Prime agricultural land is protected to give an economic boost for local farmers. Huertas, or family gardens, are encouraged. In these gardens, landowners grow vegetables, such as tomatoes and onions, and then sell them at local farmers' markets. Land is set aside for bike paths, pedestrian walkways, urban gardens, green spaces, and urban parks—all of which gives tourism a boost. Views considered scenic or historic are protected, as well.

"If you have a beautiful mountain, you should not build anything that blocks the views of that mountain," Muñoz-Criado said. "That mountain is part of the identity of that place."

The leadership of Valencia, Spain, proves that protecting and enhancing green infrastructure does not hurt economic development—it enhances it.

Cumulative Total Acres of Farmland Preserved in Maryland, 1970 to 2019

Maryland, with its combination of dense population and proximity to the waters of the Chesapeake Bay, has long been a leader in state land preservation programs. Whether it's natural lands, recreational lands, or working agricultural lands, the goal is a higher, healthier, and better quality of life for all.

We must work together to protect and restore the green infrastructure of our earth if our grandchildren are going to enjoy life on this rapidly warming and ever-more crowded planet. Protecting and restoring our green infrastructure is a large and critically important part of what it will take to get to a point where we are drawing more carbon out of the atmosphere every day than we are pumping into it. In fact, a significant percentage of the actions needed to reverse climate change involves land use, natural land restoration, and the preservation of our remaining natural lands.

The good news is that we have never had better technology for understanding where this must be done. We have never had better tools for modeling a more sustainable future. We have never had a clearer understanding of the actions necessary to bring this vision into reality. And we have never had a more urgent reason to try.

"Come, my friends, 'tis not too late to seek a newer world."

—Alfred, Lord Tennyson, "Ulysses"

Learn & Explore

Designing and Creating a Green Infrastructure
Watch a video of Arancha Muñoz-Criado's keynote presentation at the 2016 Esri International User Conference.

Protecting Urban Forests the Modern Way
See how Washington, DC, uses the ArcGIS platform to improve street tree management.

Park and Recreation Locator
This app can be used by parks and natural resources departments to help the public to discover recreational activities.

For links to these and other examples, exercises, and resources, visit SmarterGovernment.com and click chapter 11.

GrowthPrint was developed using GIS to map state programs targeting resources within priority funding areas. GrowthPrint gives a geographic representation of where smart growth makes the most sense.

12

Protecting Our Air

Today, from the perspective of space, we can see that the air we need to breathe is actually held within a very thin membrane around the earth.

This thin membrane not only holds the oxygen we need to survive but also the carbon and the methane pollution we have emitted from our burning.

Humanity has caused climate change; together, we can reverse it.

"How we treat the earth affects all people. And how
we treat one another is reflected in how we treat the earth."
—Native American proverb

A Fast-Growing Consensus on Climate Change

For the first time, according to a recent Gallup poll, a majority of Americans in every age cohort now believes that climate change is caused by human activity. Among Americans under the age of thirty, 85 percent now believe that climate change is caused by human behavior.

If you want to know where a nation is headed, talk to its young men and women—you won't find many climate change deniers among them. The vast majority believe we should be doing more, not less, to confront this dire problem.

This sort of understanding is not a passing fashion of youth. It is a strong and growing consensus. And as Thomas Paine said, "The mind once enlightened cannot again become dark."

Rethinking Our Energy Grid

Climate change is the greatest business opportunity to come to the United States in one hundred years. But this moment does not belong to the United States alone. A profound shift is underway. Other nations like Germany and China have contributed greatly to accelerating this shift in recent years.

Renewable energy technologies are improving fast. Hand in hand with those improvements are huge reductions in cost. Today, building new solar and wind generation is cheaper than building new coal or nuclear plants. And global energy markets are quickly moving to this new clean energy future.

The winning fuels of this new energy future have already been determined. By 2030, renewable energy—primarily solar, wind, and hydroelectric—will be the world's primary power source.

We can now foresee the future—a renewable energy future.

It is a future of greater prosperity, greater health, and greater well-being—not just for the United States, but for all humanity. And although some will cling to the declining profits of a failing past, smart money and smart people will invest where the future is going.

To reverse global warming, we must begin to draw down more carbon from the atmosphere than we emit. Simply leveling off carbon emissions is not enough. We must also redesign our many traditional ways in which our production of food, our built environment, our waste management, and our land use have contributed to carbon emissions. But it is entirely possible with existing technologies to change these systems so that we can draw down more carbon out of the atmosphere every day than we pump into it. And moving to a 100 percent renewable energy future is a critically important part of the larger battle. In 2015, I became the first candidate for president in either major party to propose a 100 percent clean energy grid for the United States by 2050.

Our destination is clear. Getting there quickly enough will require intention and follow-through. It will also require both a larger consensus and a new way of governing.

Our Evolving Stories and the Great Opportunity for Humanity

Unlike any other animal species on the planet, we have progressed—generationally—by cooperating with one another, by manipulating our environment (extracting resources from the soil, the waters, and the ground), and by learning to change matter to energy.

Over the last millennia or so, we human beings learned to change matter to energy primarily by means of fire. The great discovery.

Total System Change: Horatio Nelson Jackson

Horatio Nelson Jackson was a doctor from Vermont at a medical convention in San Francisco, California, in 1903.

Out to dinner with a group of other doctors, they were debating the impact of a new invention—the automobile—on American society. The other doctors thought that the automobile would have little impact—that it would have little practical use because of cost and lack of infrastructure. Horatio thought otherwise—that it would be a game changer, and that it would have a huge impact on America. He bet them all $50 that he could drive across the United States—in ninety days.

Horatio Nelson Jackson on the road in 1903. At the time, there were only 150 miles of paved roads in the US and no state highway departments. Just twenty years later, there were thousands of miles of paved roads and every state in the lower 48 had a highway department. Total transformation.

The very next day Horatio bought his first car, a Winton. He convinced a young mechanic at the car dealership, Sewall Crocker, to go with him. And they bought a dog, named Bud. The second day after the bet, Horatio, Sewall, and Bud were on their way. Remember, it was 1903. There were just eight thousand cars in the United States. Only 150 miles of paved roads. And no highway department in any of the states. Sixty-three days later—after many adventures—they drove down Fifth Avenue in New York City. They were the first people to drive across the United States in an automobile. Horatio won the $50, and he also won the bet: America did change.

Twenty years later, in 1923, there were eight million cars, hundreds of thousands of miles of paved roads, and a highway department in every state of the continental United States.

Society, now, is where Horatio was then—looking ahead at a world turned upside down. There are points in history where entire systems change. We are fast moving into just such a period now—a Third Industrial Revolution—a coming together of advanced technologies in energy, transportation, and communications that transform our world for the better.

The impetus and imperative for accelerating this change in human behavior is the existential threat of climate change and the accompanying great opportunity it portends for human life on our planet.

Fire—the burning of wood in caves. Fire—the burning of wood in stoves. And in more recent times, fire—the burning of fossil fuels. For heat, for electricity, for mobility.

In earlier generations, we thought the land, the waters, and the atmosphere were limitless, boundless things. Only recently have we come to think about the finite limits of our planet. Only recently have we come to understand the natural capacity of Earth for meeting human needs—food, water, and air.

For most of recorded human history, our stories of heroic people or individuals revolved around our capacity to conquer, our capacity to fight and to love, and our ability to overcome great suffering, sacrifice, and loss. Earth itself was merely a backdrop. Or perhaps, a wild and sometimes sinister thing to be mastered, harnessed, and exploited.

Now we find ourselves at the threshold of a new era. The emergence of what American conservationist Aldo Leopold called "The Third Ethic" (after the Decalogue and the Golden Rule):

The ethic of the land.

The ethic of the earth.

The dawn of a new consciousness.

This new ethic is greatly accelerated by a few important firsts: For the first time in human history, we have now seen ourselves from outer space as we truly are—a tiny blue capsule of life hurtling through a vast universe.

As of 2012—and also for the first time in history—a majority of human beings now lives in cities. This has never happened before in the long history of humanity.

And finally, the collective impact of our fires—the exponential growth of our traditional ways of converting matter to energy—have now heated the thin atmosphere of our planet so that we face a very real and present existential threat.

We are now pumping so much carbon and methane into the atmosphere that we are melting the ice caps of space-capsule Earth and thereby threatening its ability to support human life.

All of these are the threads of a new story. Not a story of scarcity and planetary disaster. Not a story of ruinous growth and consumption run amok. But rather, a story of greater health and well-being, greater security, and greater opportunity.

It is a story that calls upon every person, every city, every state, and every nation to pursue different and more life-giving actions.

Let me show you what I mean.

Creating Maryland's Clean Energy Future

As one state at the dawn of this new understanding, we in Maryland set three strategic goals in our pursuit of cleaner air and a more sustainable energy future. The three goals were interconnected and mutually supporting of one another. Given the accelerated pace of climate action across the globe, some of these goals in hindsight look almost quaint. At the time, many of them were considered ambitious or even stretch goals.

We set these goals in 2007:

- Reduce electricity consumption by 15 percent by 2015.
- Increase the amount of electric power coming from renewable energy sources to 20 percent by 2022.
- Reduce greenhouse gas emissions by 25 percent by 2025.

We collaboratively developed delivery plans, with shorter monthly, quarterly, and annual goals underneath each strategic goal so that everyone could see whether the actions we were taking were sufficient to reach our three strategic goals.

Cities Are Fast Going Carbon Neutral

A majority of humans now live in cities. Combine this population shift with the lethal heating of Earth's atmosphere, and one sees that *urbanization* and *climate action* have been joined together in one urgent movement of human development.

If, as the science and models tell us, 70 percent of the causes of global warming come from cities, perhaps the good news is 70 percent or more of the solutions might come from cities, as well. And well-governed and well-led cities can move much faster than states and nations.

In a bold thrust of city leadership, Helsinki, Finland, recently announced an accelerated action plan for achieving carbon neutrality by 2035. Across the key sectors of power generation, heating, transportation, waste management, and green building, the City of Helsinki has rolled out 143 leading actions to drive the city to carbon neutrality by 2035. The different choices and actions of this collaboratively developed plan seek to reduce carbon emissions by 80 percent while offsetting the remaining 20 percent with the help of carbon sinks in Helsinki and elsewhere in the world.

The solutions are not fanciful dreams. On the contrary, the solutions come from existing and proven technologies and practices. Dial up the volume of these 143 leading actions, and drive the trajectories to the goal. And all these actions can be mapped and measured.

Helsinki is a leader. But Helsinki is not alone.

The Carbon Neutral Cities Alliance is a global consortium of fast-acting cities from Boston to Berlin, from Melbourne to Vancouver, from London to Yokohama, which are all taking action to move their cities forward quickly to carbon neutrality by 2050. Along a continuum of interim to long-term goals for greenhouse gas reductions, these cities are creating a new pathway from fossil-free, to 100 percent renewable energy, to carbon neutrality.

As police officers learn best from other police officers and teachers learn best from other teachers, mayors learn best from other mayors. In city hall conference rooms around the world, mayors are now asking their teams the practical and operative question when it comes to achieving carbon neutrality—"Show me where else this is working?"

Collaboration is the new competition. Some cities will get there before others.

But the absolute direction is clear.

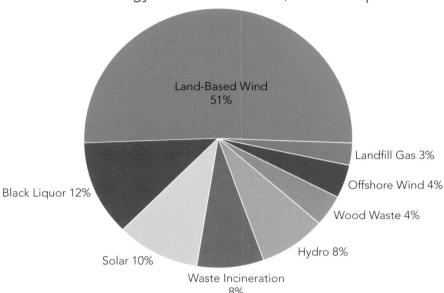

Renewable Energy Portfolio Standard, 2020 Compliance

- Land-Based Wind 51%
- Landfill Gas 3%
- Offshore Wind 4%
- Wood Waste 4%
- Hydro 8%
- Waste Incineration 8%
- Solar 10%
- Black Liquor 12%

This chart from 2012 presented our best estimate of how we could achieve 20 percent renewable energy supply in Maryland by 2020. With the cost of renewable energy now less expensive than fossil fuels, states are adopting higher and higher goals in the movement toward 100 percent renewable electricity.

Small things done well make bigger things possible.

If the difference between a dream and a goal is a deadline, then the difference between hitting your goal or not is the sufficiency of your leading actions—or more specifically, a host of leading actions, often referred to in government speak as "programs." In this case, programs designed to bring about different outcomes in energy conservation, greenhouse gas reductions, and renewable energy production.

By 2015, through the combination of many leading actions, we were driving toward each of these mutually interdependent goals. Smart meters, resilient grids, home energy audits, participation in a cap-and-trade regimen known as the Regional Greenhouse Gas Initiative (RGGI), solar incentives, offshore wind, clean-air retrofits at power plants, electric car incentives, energy retrofits for homes and businesses, green buildings—all these things, and more, were working together to drive us toward achieving our strategic goals by the deadlines we had set.

Reducing Energy Consumption

Our first goal was to reduce both per capita peak demand and per capita electricity consumption by 15 percent by 2015. Toward this goal, the Maryland General Assembly passed the EmPOWER Maryland Energy Efficiency Act in 2008. This legislation set a target reduction of 15 percent from a 2007 baseline in per capita electricity consumption and peak demand by 2015.

Under the EmPOWER Maryland Energy Efficiency Act, programs were established to mandate and incentivize utilities to drive leading actions in residential, commercial, and industrial energy efficiency. The actions under each program were divided between demand response, and energy efficiency and conservation. EmPOWER programs began in 2009, with a second round of planning and approvals in the fall of 2011. The final Public Service Commission decisions from a third round of planning happened in 2015. In reality, the adaptation and constant improvements should never stop.

In the context of EmPOWER, demand response programs were voluntary and primarily designed to reduce the amount of power needed, especially during times of peak demand. The EmPOWER Maryland Energy Efficiency Act required utilities to implement cost-effective demand response programs designed to achieve a reduction in their per capita peak energy demand of 5 percent by 2011, 10 percent by 2013, and 15 percent by 2015.

Our energy efficiency and conservation programs were primarily designed to reduce the amount of energy consumed throughout the year. Residential energy efficiency and conservation programs included appliance, HVAC, and lighting rebates; home performance with ENERGY STAR; and quick home energy checkups. For commercial and industrial customers, utilities offered lighting and equipment rebates, retro-commissioning, and rebates for custom projects.

Some of our most successful leading actions included:

- The Quick Home Energy Check-Up Program, in which an energy professional performs a walkthrough to help the homeowner identify opportunities to reduce energy use, including checking insulation, air leakage, heating and cooling systems, windows and doors, lighting and appliances, water heating equipment, and more.
- Low-Income Energy Efficiency Programs, which help low-income households with installation of materials and equipment to reduce energy use at no charge.
- Project Sunburst, which provided government entities with grants that funded the installation of solar arrays on state buildings.
- The High-Performance Building Law, which required new or significantly renovated state facilities to achieve at least a Silver rating as determined by the US Green Building Council's LEED rating system.
- The Freedom Fleet Voucher Program, which provided financial assistance for the purchase of new and converted alternative fueled-vehicles registered in the State of Maryland.
- The Game Changer Program, designed to mitigate the additional costs and risks of installing game-changing technologies, evaluate the efficacy of those technologies through performance data

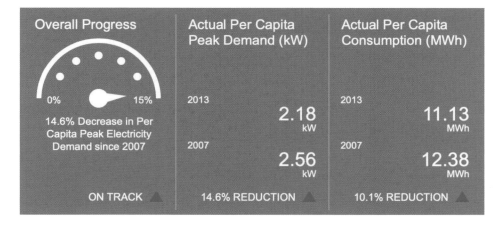

A StateStat dashboard showing progress towards the state strategic goal of reducing electricity consumption by 15 percent by 2015 (as of July 1, 2014).

collection and analysis, and assess the cost/benefit ratios of those technologies through life-cycle analyses.
- Net metering legislation, which changed the net energy metering program for solar and other energy producers to more easily allow small-scale generators, such as rooftop solar systems, to be compensated when they produced more energy than they consumed.

The effectiveness of every individual leading action or program to reduce energy consumption could be judged by the degree to which it contributed to the trajectory of progress to our goal. As years progressed and we learned from experience, we sought to do more of whatever worked best to help us achieve our goal.

As a result of programs offered by Maryland's utilities, per capita peak demand was decreased by 14.6 percent and per capita consumption was decreased by 10.1 percent from 2007 to 2015.

EmPOWER Maryland programs saved more than $3.7 billion in lifetime energy costs, and more than thirty-seven thousand commercial and industrial businesses participated. The programs were so successful that Maryland was ranked in the top ten states for energy efficiency over a three-year period, after being ranked twentieth in 2006.

In 2007, Maryland had 0.1 megawatts of solar capacity on its grid. By 2013—with policy to encourage the move to renewables—we added 173 megawatts of solar. And we can't stop there.

As a result of the Quick Home Energy Check-Up Program, more than 466,000 energy-efficient appliances and HVAC systems had been installed in more than 27,502 Maryland households by November 2014.

As of November 2014, more than 29,000 low- and moderate-income Maryland households had benefitted from energy efficiency retrofits through state-administered Low-Income Energy Efficiency Programs.

The nearly 10 megawatts of energy produced by Project Sunburst almost tripled the amount of solar energy on Maryland's grid, and helped the state reduce government expenses by lowering long-term electricity bills.

Between 2007 and 2015, forty-two new fully state-funded buildings were designed to comply with the Maryland Green Building Council's High-Performance Building Program, and seventy-eight schools were also on track to comply with the program.

In 2013 alone, 2,253 residential and commercial grants were issued as part of the Game Changer Program, helping Marylanders save more than $37.7 million in energy costs over fifteen years.

Increasing Renewable Energy Generation

In 2007, we also set the strategic goal of producing 20 percent of all electricity generated within Maryland from renewable sources by 2022. When the goal was initially created, only 5.8 percent of total energy generation came from renewable energy, totaling 861 megawatts. By 2015, the state's capacity had increased to 8.2 percent (1,032 megawatts) and future projections showed that we were on track to meet the 2022 goal.

Here's how.

To reach the 2022 goal, we enacted the Renewable Portfolio Standard (RPS), a statute that required 20 percent of all electricity sales in Maryland to be from renewable energy, as evidenced by the production or purchase of renewable energy credits. Maryland's strategy for meeting the in-state generation goal included a mix of renewable resources, from distributed solar generation to offshore wind.

Leading actions in pursuit of our strategy included the following:

- The RPS carved out 2 percent, or about 1,200 megawatts, of the state's energy to come from solar by 2020. As a result of available incentives, grants, and tax credits, choosing solar became much easier for residents and businesses in the state.
- To help increase our ability to meet RPS goals, we passed a bill that made Maryland the first state in the country to allow the energy generated by geothermal heating and cooling technologies to be eligible for RPS as a Tier 1 renewable source. This was on top of our residential and commercial geothermal grant programs. According to the US Environmental Protection Agency (EPA), geothermal heating and cooling systems are the cleanest, most energy-efficient, and most cost-effective space conditioning systems available.
- Biomass is organic matter from plants or animals that can be used as a fuel, normally in a power station to generate electricity. In 2012, we produced a biomass bill that incorporated generation facilities powered by animal waste for thermal applications as a Tier 1 resource in Maryland's Renewable Portfolio Standard. To support the progress of biomass in the state, Maryland initiated an E85 (an ethanol fuel blend of 85 percent denatured ethanol fuel and 15 percent gasoline) refueling station program with sixteen fueling stations—the number of stations has now expanded exponentially.
- Through the Generating Clean Horizons Program and a partnership with the University of Maryland,

we increased our land-based wind energy generation. In 2014, we expanded the availability of wind grants from solely business, residential, and non-profits to community-based projects.

- In 2013, we passed the Maryland Offshore Wind Energy Act, creating new incentives for up to 500 megawatts of offshore wind capacity off Maryland's Atlantic Coast. Our efforts to spur deployment of offshore wind energy focused on working with the US Department of the Interior to ensure proper leasing of the Maryland Wind Energy Area, developing strategies and partnerships with other entities, and utilizing the Offshore Wind Development Fund to gather data and invest in infrastructure that would drive down the cost of offshore wind energy.

Experience was our best teacher. We looked to expand as quickly as we could the actions that were producing the best results for the buck. Here are some of the things we accomplished and learned in Maryland:

- In 2007, Maryland had 0.1 megawatts of solar capacity on its grid, and we were able to grow that to 173 megawatts by 2013.
- A study by the Department of the Environment's SunShot Program found that Maryland created 2,000 solar industry jobs, making itself a leader among states in solar energy jobs.
- The state's thirty-five solar school programs alone generated 9,485,605 kilowatts of solar capacity annually, or $1,015,168 in savings.
- Under the geothermal heating and cooling grant programs, we incentivized 3,421 residential and fourteen commercial installations, resulting in 15,726,060 KWh saved.
- E85 gasoline purchases were increased in Maryland from 24,448 gallons in the first quarter of 2008 to 66,124 gallons in the first quarter of 2013.
- Biomass-sourced electricity in Maryland was increased from 200 megawatts in the 1980s to more than 8,000 megawatts by 2015, a 4,000 percent increase.
- In 2007, Maryland produced zero megawatts of energy from land-based wind. As of 2014, annual generation from new land-based wind was increased to 314,572 megawatts.
- In August 2014, Maryland successfully auctioned off nearly 80,000 acres for future offshore wind development off Maryland's Atlantic Coast.

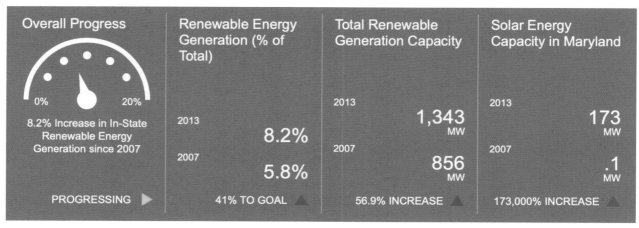

Overall Progress	Renewable Energy Generation (% of Total)	Total Renewable Generation Capacity	Solar Energy Capacity in Maryland
0% 20%		2013	2013
8.2% Increase in In-State Renewable Energy Generation since 2007	2013 **8.2%**	**1,343** MW	**173** MW
	2007 **5.8%**	2007 **856** MW	2007 **.1** MW
PROGRESSING ▶	41% TO GOAL ▲	56.9% INCREASE ▲	173,000% INCREASE ▲

A StateStat dashboard showing progress towards the state strategic goal of increasing Maryland's renewable energy generation to 20 percent of its portfolio by 2022 (as of July 1, 2014).

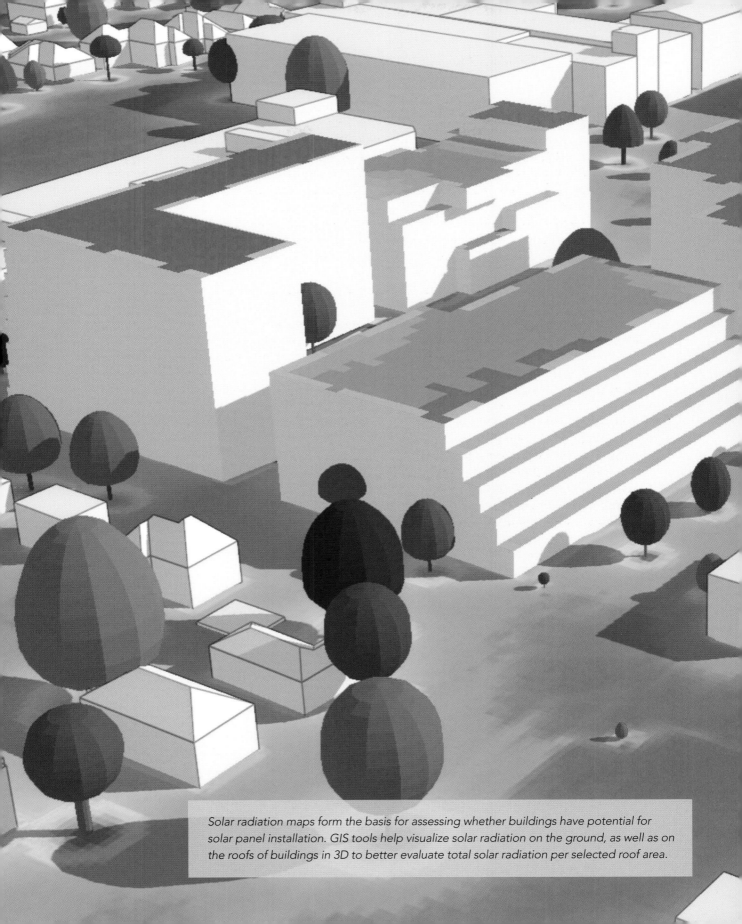

Solar radiation maps form the basis for assessing whether buildings have potential for solar panel installation. GIS tools help visualize solar radiation on the ground, as well as on the roofs of buildings in 3D to better evaluate total solar radiation per selected roof area.

Project Drawdown

There's more than one way to lead the fight against climate change. In fact, there are more than one hundred ways.

Globally, half of all investment in new energy production is being invested into the development of renewable energy. Green building technology and adoption is rapidly spreading and becoming more and more commonplace. And recently, my own state of Maryland gave a unanimous go-ahead for the largest off-shore wind farm ever constructed off the US Atlantic coast.

Humanity has reached a vital inflection point of understanding. It is possible with existing technology and science to reverse and draw down the amount of carbon dioxide we humans pump into the air. "Project Drawdown" shows us how.

Drawdown: The Most Comprehensive Plan Ever Proposed to Reverse Global Warming has been led, inspired, and edited by Paul Hawken—a man I am honored to call a friend. Paul is one of the most fearless and inspiring American leaders in the world today on the great challenge of climate change. His *magnum opus* is a research project of global proportions. Drawing from the best science on best practices, this best-selling book— published by Penguin in twelve different languages—is a crowd-sourced healing of the deepest and potentially most far-reaching kind.

Don't let the words "book," "research," "or "science" fool you. This book is not another exercise in connecting the dots of physics into a straight line that leads us all to hell. Paul brings to this project the "can-do American pragmatism"—and belief—that created the American Revolution, won world wars, and landed a man on the moon. It has been my honor to serve with many others on the advisory board of this critically needed work.

Most people with a basic understanding of the problem of air pollution can see that it is not enough to level off emissions. We must actually redesign our human activities on Earth's surface so that we draw more carbon and other harmful emissions out of the atmosphere every year than we pump into it.
And we can.

Drawdown details the 100 most important solutions for combatting global warming— their history, carbon impact, costs and savings, pathways to adoption, and how they work. By mapping, measuring, modeling, and describing these solutions, *Drawdown* aims to reverse the accumulation of carbon in the atmosphere in the next thirty years.

And the best news is that all these solutions are already well-known and understood, scientifically proven, and established—and expanding—around the world.

The practicality of *Drawdown* is an important factor in its potential for success. It lays out two general types of solutions that can bring down carbon concentration in the atmosphere:

1. Those that reduce carbon emissions (such as increasing the use of geothermal energy and the adoption of plant-rich diets)
2. Those that help the atmosphere absorb more CO_2 (such as better managing grazing and planting more trees)

The book then lays out three scenarios for global adoption of the solutions:

1. In the "plausible scenario," the solutions are adopted at a realistic yet optimistic rate.
2. In the "drawdown scenario," the solutions are adopted at a rate that achieves drawdown by 2050.
3. In the "optimum scenario," the solutions are adopted at their maximum potential.

Wind energy, solar farms, silvopasture, education, green buildings, new refrigeration technology—each of these actions, based on current technology, is ranked by scientific criteria from most impactful to least. The cost of creating a more prosperous, healthier, and more secure future for our kids turns out to be less than the cost of clinging to business as usual. That's right—creating a more sustainable way of living turns out to be more profitable than pumping more carbon into our thin atmosphere.

And although the book previews upcoming technologies on the horizon such as hyperloop transportation and other promising innovations on the energy frontier, the real service of this great work is understanding what is actually possible, here and now, *with existing technologies and science.*

Reading through the pages and pictures of *Drawdown* with its clear, plain-spoken explanations, photos, and graphics, one feels lifted up by the wellspring of goodness and humanity that has been brought together in these life-giving pages.

We need not let the darkness of denial rob us of our ability to see, to feel, and to act. To quote Thomas Paine, "We have it in our power to begin the world over again."

Project Drawdown shows us how.

Reducing Greenhouse Gases

In 2007, we also set an aggressive goal of reducing greenhouse gas emissions in the state by 25 percent by 2020 and implemented one of the most proactive greenhouse gas emission programs in the nation.

To achieve our 2020 goal, the Maryland Commission on Climate Change was created and tasked with developing a Climate Action Plan. The plan encompassed 150-plus leading actions (programs and initiatives) to reduce Maryland's greenhouse gas emissions by 55 million metric tons of carbon dioxide-equivalent (MTCO2e) annually.

To achieve this goal, a host of leading actions were implemented.

In the Regional Greenhouse Gas Initiative, Maryland joined with nine states to create the nation's first regional cap-and-trade program for carbon dioxide that puts a price on the carbon that power plants dump into the atmosphere. We calculated that Maryland's participation in this program would result in 3.6 million metric tons of carbon dioxide (CO_2) reductions by 2020. To support the state CO_2 Budget Trading Programs, states participating in this initiative implemented a regional auction platform to sell CO_2 allowances.

Regenerative Neighborhoods: Project Drawdown in Action

When it comes to mapping out the future, it's not such a colossal task that only government bureaucracy can attempt. In many places, nonprofits, businesses, and individuals are stepping up to tackle big issues at much more granular scales.

In West Baltimore, architects Michael Hindle and Carri Beer have proposed a "Regenerative Neighborhood" redevelopment plan in the hardest-hit areas. These urban areas have been reduced to a wasteland, and the architects have somewhat free rein to establish a systems approach to optimizing the interconnected aspects of regeneration. The goal is to use holistic, regenerative design to leverage value in places like West Baltimore, where banks and conventional development teams do not see value, all while protecting existing residents against gentrification.

Using Project Drawdown as a scorecard, Hindle and Beer have identified a total of thirty-eight out of the one hundred solutions to reverse global warming that are covered in the book—from rooftop solar to biomass and methane digesters; family planning, education, and walkability; electric vehicles, ridesharing, and much more—that can be integrated into their Regenerative Neighborhood project and implemented at the neighborhood level.

This example of tackling global problems at neighborhood scale shows that we need not wait for national policy makers to take actions that matter.

As required by the Greenhouse Gas Reduction Act of 2009, the Maryland Department of the Environment conducts a full inventory of greenhouse gas emissions every three years.

The Maryland Clean Cars legislation adopted stricter vehicle emission standards and directly regulated carbon dioxide emissions. These standards became effective in Maryland for model year 2011 vehicles, significantly reducing a number of emissions, including volatile organic compounds and nitrogen oxide. The Clean Cars Program included a Zero Emissions Vehicle mandate that car manufacturers must meet. Maryland joined twelve other states in a pledge to bring 3.3 million Zero Emissions Vehicles on the road by 2025. The state also developed the Freedom Fleet Voucher Program to provide financial assistance for the purchase of new and converted alternative-fueled vehicles registered in Maryland.

Maryland passed the Healthy Air Act, which required major reductions in air pollutants to be phased in at Maryland power plants starting in 2009, with additional reductions in 2012 and 2013. At full implementation, the Healthy Air Act was designed to reduce nitrogen oxide emissions by approximately 75 percent and sulfur oxide emissions by approximately 85 percent from 2002 levels.

Our Zero Waste initiative called for the near elimination of solid waste sent to landfills or incinerators for disposal. Instead, the vast majority of Maryland's solid waste was planned to be reused, recycled, composted, or prevented through source reduction.

Maryland established tree-planting goals as part of both the state-specific Greenhouse Gas Reduction Act carbon reductions to be achieved by 2020, and as part of the EPA -mandated Watershed Implementation Plan to achieve nitrogen, phosphorus, and sediment reductions from our waters by 2025.

And because managing forests sustainably captures carbon, we promoted sustainable management practices in existing Maryland forests on both public and private lands. We set a goal to achieve the afforestation and/or reforestation of 43,030 acres in Maryland by 2020.

The effectiveness of every individual leading action or program to reduce greenhouse gas emissions could be judged by the degree to which it contributed to the trajectory of progress to our goal. We never stopped striving and searching for the actions that were most effective in driving us to our goal. Here are some of the things we accomplished and learned:

- By 2015, Maryland had reduced overall emissions from a 2006 baseline by 9.7 percent.
- By 2015, more than $3 million had been added to Maryland's economy due to the regional cap-and-

A StateStat dashboard showing progress toward the state strategic goal of reducing Maryland's greenhouse gas emissions by 25 percent by 2020 (as of July 1, 2014).

trade program created under the Regional Greenhouse Gas Initiative.

- The twenty-fifth auction of CO2 allowances was held on September 3, 2014. Maryland, as a participating state, sold 3,725,943 allowances for year 2014 at $4.88 per allowance, resulting in over $1 billion in proceeds.
- By 2015, more than three thousand electric vehicles had been registered in Maryland.
- As of 2013, the Freedom Fleet Voucher Program had made fifty awards, displacing approximately 585,000 gallons of fuel annually.
- As of 2013, more than 10,000 tons of nitrogen oxide and over twenty-three thousand tons of sulfur oxide had been driven down by the Healthy Air Act.
- In 2012, Maryland's source reduction and recycling activities reduced greenhouse gas emissions by 6.5 million MTCO2e, relative to no source reduction or recycling.
- Maryland inmates contributed a million tree seedlings along with the Department of Natural Resources.
- Between 2006 and 2015, Maryland achieved a 39 percent carbon reduction goal under its Greenhouse Gas Reduction Act.

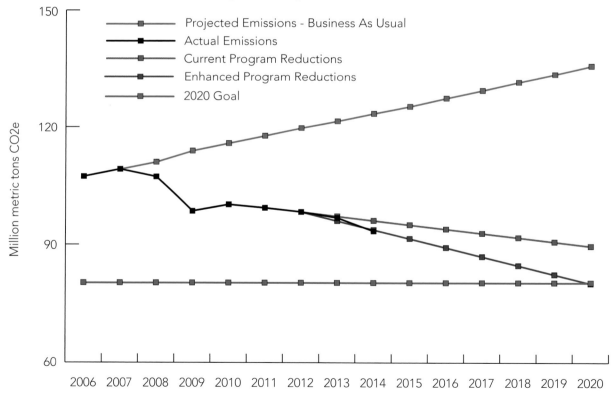

To achieve a 25 percent reduction in Maryland's greenhouse gas emissions by 2020, we implemented more than 150 programs and initiatives as described in the delivery plan for achieving this important goal. The reduction included offsetting the projected increase that was otherwise expected to occur between 2006 and 2020 (called the 'business as usual' forecast).

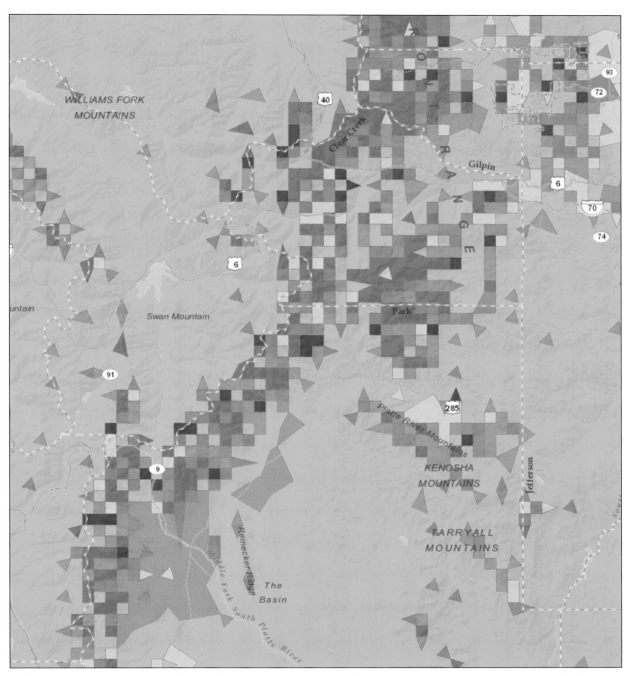

Selecting an appropriate site is key to the success of any renewable energy project. The site plays a crucial role in financial returns and ease of construction, as well as to ongoing operations, maintenance, and overall safety. Maps of wind power potential, such as this one covering an area southwest of Denver, Colorado, are used by planners to identify new sites for the installation of high-efficiency wind turbines.

America's Renewable Energy Future

The policy choices that accelerate a shift to renewable energy in the United States are very clear. These are the choices, broadly speaking:

1. Realign utility profit incentives away from increased energy consumption, and forward to the maintenance of a secure, reliable, and dynamic grid where "prosumer customers" produce energy as well as consume it.
2. Shift subsidies and incentives away from fossil fuels and toward renewables.
3. Invest in research and development that drive America's renewable energy future.

Historically in our country, it is states and cities that drive energy policy, not the White House. It would be good to have the White House leading this movement instead of trying to put the brakes on it. But the increasing cost advantages of renewable energy and the actions that states and cities have taken to powerful effect means more rapid progress is possible, even now.

In fact, all the Paris commitments for emissions reductions in the United States—that is, a reduction in greenhouse gas emissions of 26 to 28 percent below 2005 levels by 2025—can be met by the combined actions of our cities and states. What are those actions?

As of 2018, nineteen states had adopted greenhouse gas reduction targets. Ten northeastern states have joined in a mandatory cap-and-trade regimen to set a price on carbon emissions, and it is working. California is large enough to have enacted its own mandatory cap-and-trade program, and it too, is working. Other states will follow. Twenty states have now adopted decoupling policies that remove the traditional utility profit incentive from increased consumption. Twenty-six states have now adopted mandatory renewable energy portfolio standards. Twenty-two states have now adopted energy efficiency standards and targets. Twenty-six states have now adopted residential building energy codes, and four states have recently adopted higher standards. Forty-one states have now adopted net-metering so homeowners can sell excess energy they produce back to the grid.

Cities and states across the country are now passing legislation and enacting ordinances to honor the Paris carbon reduction targets with action. From Los Angeles, California, to Atlanta, Georgia; from Orlando, Florida, to Somerville, Massachusetts—cities are moving to 100 percent renewable.

This movement from fossil fuel consumer to renewable energy prosumer is also being driven by increasing numbers of colleges, universities, businesses, and households.

The state of Hawaii has set a goal of 100 percent renewable energy by 2045. Given the cost of shipping fossil fuel to the island state, Hawaii might well be the first state to reach a 100 percent renewable grid, but it will not be the last. California, now the fifth largest economy in the world, is on target to hit 50 percent renewable by 2030. Texas will soon reach 25 percent renewables in its energy mix. And Iowa is fast approaching 40 percent from renewables.

In 2017, the people of New Jersey inaugurated a new governor in Phil Murphy. Murphy campaigned on moving New Jersey forward to a 100 percent renewable grid by 2050. He won by overwhelming margins—especially among the young and among people who live in cities.

In every election for state offices this year and in the years ahead, candidates will be asked where they stand on climate action. Does a candidate believe we should be accelerating the move to renewable energy or putting the brakes on it? And increasing numbers of winning candidates in both parties will answer with support for actions that move us forward, not back.

A new energy story is emerging. It is driven by technological advances, by the urgent threat of climate change, and by the unprecedented opportunity of zero-cost energy. It is driven by actions, not words.

And its progress can be measured and mapped all across the United States and the world.

Learn & Explore

Seizing America's Renewable Energy Future
Watch a video from my keynote address at the 2018 Sustainability Symposium.

Calculating Rooftop Solar Potential
Local governments can use this tool to calculate solar radiation maps for the wider community.

Understanding Ocean Wind Energy
This story map helps to identify ocean wind energy potential off the shore of the United States.

For links to these and other examples, exercises, and resources, visit SmarterGovernment.com and click chapter 12.

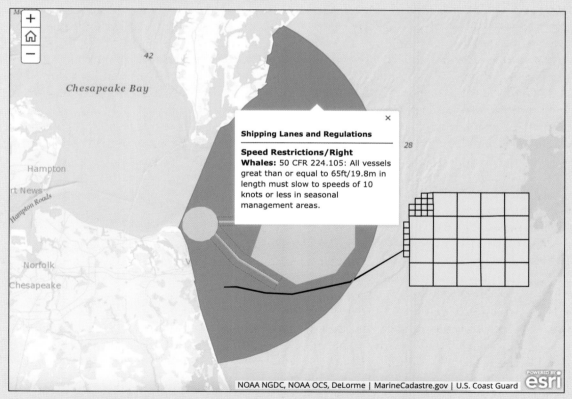

This interactive map from the Bureau of Ocean Energy Management (BOEM) and the National Oceanic and Atmospheric Administration (NOAA) identifies active renewable energy leases near Chesapeake Bay, showing where companies have leased areas of the ocean with the intent to build wind energy facilities.

The American Revolution, and Our World

Leadership is an art. Politics is a game of timing, character, and chance. But effective governance is a science.

And as a science, effective governance is subject to the universal laws of physics and math. Its choices and actions are subject to the measurements of cause and effect; its theories and practices are subject to peer review. And for a self-governing people, the by-products of effective public service are progress, legitimacy, and trust.

"If we could first know where we are, and whither we are tending, we could then better judge what to do, and how to do it."

—Abraham Lincoln

"The whole world is my village."

—Mahatma Gandhi

"A Republic, If You Can Keep It"

The story is told that at the end of the Constitutional Convention in 1787, a passerby approached Benjamin Franklin as he left Independence Hall and asked, "Well, Doctor, what have we got, a republic or a monarchy?"

Franklin answered with eyebrows raised over his spectacles, "A republic, if you can keep it."

This "garden of democracy" requires constant work and tending. Each of us is needed. To protect the security and well-being of our people, to construct an economy that serves the needs of the many, to safeguard the exercise of individual freedoms that make life worth living—these are not small things. These are big things.

And it is the ongoing pursuit of these goals, and the success of this experiment to date, that forms the beating heart of the American Revolution today—the living republic that each of us has a responsibility to keep.

The Father of American Public Administration

George Washington is popularly remembered as the father of our country— "First in war, first in peace, and first in the hearts of his countrymen." But as a matter of historic fact, George Washington was also the first public administrator of the United States. And after he was elected, President Washington was laser-focused on establishing "the legitimacy" of the new federal government in the eyes of its citizens.

President Washington understood that for the government to be legitimate, the government had to work. And for the government to work, it required effective public administrators. He believed in a merit-based civil service. He believed in a civil service that reflected the diversity of the people it served. And he understood that the legitimacy of the new republic's government would ultimately be based on

For George Washington—the first public administrator of the Republic—effective governance was a longitudinal science informed by objective evidence and experience of what works.

its effectiveness in addressing and solving problems for its citizens.

He was also a cartographer (a mapmaker) by training, and a master of spatial intelligence on and off the battlefield.

As a public administrator, Washington also believed there should be objective criteria for making policy decisions. And to this end, he was guided by certain propositions—many of which found their roots in the Scottish Enlightenment.

He believed in the rule of law. He understood the power of precedents, but he also believed that effective governance in a democracy must necessarily be a longitudinal experiment—a series of hypothesis, trial, and error, all driving towards achieving results. He called public administration "the science of government." And he believed that with reflection, benevolence, experience, and persistent effort, this experiment would advance the common good we share as a nation.

The following is the conclusion of an insightful article written by Scott A. Cook,

George Washington was also a surveyor and mapmaker by profession. A revolutionary answer to the question, "Can you show me my house?" is provided by this map he made of his own home, Mount Vernon, Virginia, in 1766.

PhD, at the Air Command and Staff College, and William Earle Klay, PhD, at Florida State University, about George Washington—the father of American public administration.

"In a sense, American public administrators today are part of a longitudinal experiment, begun by Washington, testing the staying power of a representative democracy. Perpetuating the success of that ongoing experiment is an essential task of practitioners and scholars in the United States... The conceptual framework applied by Washington encompassed both norms of behavior and the doing of objective empirical analysis."

George Washington's propositions for effective public administration, back in the day, included:

- Respect for the rule of law
- Civilian control of the military
- Accountability to the public
- An obligation of efficiency
- Merit selection in civil service
- Representativeness (diversity) in hiring and contracting
- Education for administrators in the "science of government"
- Objective policy analysis

More than two hundred years later, echoes of Washington's propositions can be heard today in the criteria used by the World Bank to assess the integrity of governments the world over.

In a modern take on what constitutes legitimacy, the World Bank writes: "Governance consists of the traditions and institutions by which authority in a country is exercised. This includes the process by which governments are selected, monitored and replaced; the capacity of the government to effectively formulate and implement sound policies; and the respect of citizens and the state for the institutions that govern economic and social interactions among them."

Like George Washington's own propositions, the World Bank has also defined a set of criteria and a coloring scheme for ranking the nations of the world as to how closely they come to living up to the promise of good governance. Those indicators are:

Individual voice and accountability. Rule of law. Regulatory quality. Control of corruption. Government effectiveness. Political stability. Absence of violence or terrorism.

The Information Age has brought about some important changes to the old ways of governing. This age has also brought changes to the way modern leaders maintain positional advantage and legitimacy of government action in the eyes of the self-governed. But the flood of new information and new technologies has also made collaborative leadership more important today than ever before.

Toward FedStat in the United States

One thing I believe I have learned in fifteen years of executive service—taking CompStat to CitiStat, taking CitiStat to StateStat, and beyond—is the larger the human organization, and the larger the area, the more important performance management and the map become.

Department of Housing and Urban Development (HUD) Secretary Shaun Donovan (center) and other senior officials conduct a quarterly performance review at a HUDStat meeting on November 10, 2010.

The framers of our Constitution never set out to create a nation that muddles through; we came together to form a more perfect union—from generation to generation. We came together to protect one another, and to give our children lives that could be lived more fully, more securely, more abundantly.

Data-driven decision-making and performance management are essential to that ongoing mission. And effective governance is critical to the success of our American experiment in democracy.

In 2015, the federal Office of Management and Budget (OMB) launched a data-driven budget review process called FedStat. This followed a number of agency-specific initiatives developed by OMB and the White House, including TechStat for troubled IT projects; PortfolioStat to bring IT and mission stakeholders together; and others, such as CyberStat, AcqStat, and HRStat. Other pioneering departmental efforts included HUDStat and NASAStat.

"We're doing a whole set of other things around the federal government around performance management," said Beth Cobert, then the deputy director for management at OMB. "Whether it's agency priority goals, agency strategic reviews that are mandated by the Government Performance and Results Modernization Act (GPRMA). How can we make these things all part of one effort? What emerged from that was FedStat."

The FedStat process involved looking at key performance indicators (KPIs) to evaluate budget development and compare one set of programs with similar programs in other agencies.

These steps toward a comprehensive FedStat performance management system at the White House were a good start. Sadly, they never got to the point of being a true performance management

Modern performance management dashboard apps are increasingly being used by government agencies to monitor key performance metrics and communicate progress made on strategic outcomes, both internally and with the public.

Worldwide Governance Indicators

Governance consists of the traditions and institutions by which authority in a country is exercised. This includes the process by which governments are selected, monitored, and replaced; the capacity of the government to effectively formulate and implement sound policies; and the respect of citizens and the state for the institutions that govern economic and social interactions among them.

The Worldwide Governance Indicators (WGI) project by the World Bank aggregates governance indicators for six dimensions of governance:

• Voice and Accountability
• Political Stability and Absence of Violence/Terrorism
• Government Effectiveness
• Regulatory Quality
• Rule of Law
• Control of Corruption

Voice and accountability captures perceptions of the extent to which a country's citizens are able to participate in selecting their government, as well as freedom of expression, freedom of association, and a free media.

Political stability and absence of violence/terrorism measures perceptions of the likelihood of political instability and/or politically-motivated violence, including terrorism.

Government effectiveness captures perceptions of the quality of public services, the quality of the civil service and the degree of its independence from political pressures, the quality of policy formulation and implementation, and the credibility of the government's commitment to such policies.

Regulatory quality captures perceptions of the ability of the government to formulate and implement sound policies and regulations that permit and promote private-sector development.

Rule of law captures perceptions of the extent to which agents have confidence in and abide by the rules of society, and in particular the quality of contract enforcement, property rights, the police, and the courts, as well as the likelihood of crime and violence.

Control of corruption captures perceptions of the extent to which public power is exercised for private gain, including both petty and grand forms of corruption, as well as "capture" of the state by elites and private interests.

enterprise—delegation of such efforts to any budget director's office usually guarantees a lack of departmental adoption or uptake across the enterprise.

The problem at our federal level in those early years of FedStat was not a lack of goals or a lack of data. Agencies had dozens of goals, and hundreds of KPIs. The problem was knowing which ones really mattered. The problem was understanding what big, strategic goals—if any—our nation was pursuing, and why.

Having executive meetings, of course, is not a goal. But in the absence of a regular cadence of accountability—driven directly by the chief executive, not his budget director—it was difficult to tell which federal goals were the most important and which actions were the most impactful for realizing them.

A notable and hard-fought exception to this norm was the Affordable Care Act—even with its rocky implementation. Another exception was the administration of the American Recovery and Reinvestment Act—a shining, if little praised, example of a new way of governing—the efficacy of the data, the map, and the method.

Most of the federal goals of this earliest incarnation of FedStat, however, were about efficiency targets, process, and inputs, rather than meaningful results and outcomes. And all this process meant very, very little to a citizenry grown weary of declining wages and a long recession.

At our national level, we must develop a clearer view of the most important things our government is trying to accomplish and why. This view requires not only a national story for winning this Third Industrial Revolution, but clear and measurable goals that advance us to victory.

If we are going to make our children winners in this changing economy, Americans should know what our federal government's top five objectives are:

1. Job creation.
2. Improving the security of our people.
3. Improving the education and skills of our people.
4. Improving the sustainability of our way of life.
5. Improving the health and well-being of all Americans.

Federal employees should know how their work and their particular agency contribute to the achievement of these objectives. And leaders, staff, and all of us as citizens should all know whether we're making progress together, or where greater work and greater effort still remain to be done.

The first step in implementing any stat process is opening the data to the public. During President Barack Obama's honorable years of service, the executive order of "open data" was clearly and repeatedly stated. The Obama Administration made real strides toward making that data understandable for citizens and actionable for public administrators. And for these strides, his leadership should be applauded.

But just days after President Obama left office, the website of the White House open data portal was taken down. At the time of this writing, it remains dark.

The World as a Village

The United Nations (UN) has recently undertaken the challenge of helping countries achieve important social, economic, and environmental goals by 2030. Known as the Sustainable Development Goals (SDGs), these seventeen measurable objectives range from eradicating poverty, ending hunger, and ensuring gender equality to building modern infrastructure and creating inclusive and sustainable cities. Some of the other goals include protecting land, conserving water, and accelerating the shift to renewable energy.

The SDGs follow on the success of the UN-backed Millennium Development Goals, which drew to a close in 2015. Pursuing the Millennium Development Goals lifted millions of people out of poverty and saved the lives of hundreds of thousands of men, women, and children from deadly diseases.

The initial recommendation to the UN was for just ten SDGs. In the collaborative, international process that followed, these ten goals grew to seventeen.

Like the living systems of our earth, the SDGs are interconnected, interdependent, and holistic. "One of the fundamental philosophies behind the SDGs is that development is not [done] by segments and areas," Stefan Schweinfest, director of the United Nations Statistics Division (UNSD), said in a UN Web TV video interview in 2017. "You cannot just develop the education sector without paying attention to the health sector. If kids are sick, they will not go to school, even if you have the best schools in the world. We have to [address dependencies like these] with integrated data systems, where we can also relate the various information systems to each other."

For the first time, the United Nations Statistical Commission is implementing a globally agreed-upon framework for each country to measure its progress in an open, timely, and accurate manner—shared by all. The purpose is to encourage member states to take actions based on data-driven decisions to make positive change at a faster pace.

It's like CitiStat for the world. The mission is larger, the platform is larger, and so are the stakes. But the management principles are the same. It is a matter of combining geospatial and statistical information to visualize patterns, address data gaps, build models, and effectively target resources to areas demonstrating the most need in order to improve sustainable development outcomes overall.

Timely, accurate information shared by all. Rapid deployment of resources. Effective tactics and strategies.

The United Nations Sustainable Development Goals are designed to guide countries around the globe in achieving critically important social, economic, and environmental goals for human living by 2030.

Place matters, and location—anywhere on the surface of the planet—is the universal key to integrating information about society, the economy, and the environment. Location is also the key to measuring the effectiveness of actions (the tactics and strategies) undertaken to achieve the SDGs.

This is not merely an exercise in analysis, but rather analysis in service of taking more life-giving actions the world over. And this is exactly what the new web mapping and data management platform—called the Federated Information System for the SDGs—will do.

Organizing Data around Goals and Leading Actions

Stefan Schweinfest discussed the Federated Information System for the SDGs, also known as the SDG Hub, at the Geospatial World Forum in 2017. The SDGs are based on the 2030 Agenda for Sustainable Development. In order to achieve the SDGs, nations must overcome significant obstacles to wiping out poverty, tackling injustice, and healing environmental damage. The ambitious and transformative agenda encourages member states to measure, monitor, and drive progress with actions based on the best available evidence and science.

Action is the driver. Geography is the common platform. Technology and data are the universal enablers for the feedback loops of action that tell us whether or not we are making progress toward achieving our goals.

The Agenda for Sustainable Development is not a dream; it is a goal. And the deadline is 2030.

Each SDG is composed of specific, action-oriented sub-targets. The 2030 agenda identifies and defines

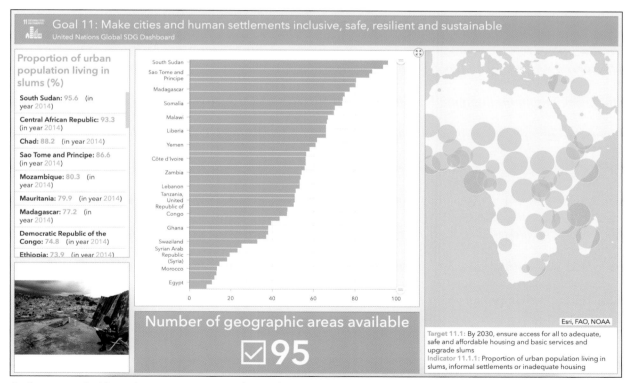

Performance dashboards give nation-states the tools to monitor and drive progress toward achieving the United Nations Sustainable Development Goals by 2030. The difference between a dream and a goal is a deadline.

a total of 169 sub-targets. The targets, in turn, have their own indicators that measure the leading actions required to drive progress. Currently, these global leading actions are tallied at 232.

Individual countries must set their own national and regional targets. They must develop their own action plans in light of their unique challenges, natural resources, capacities, and strengths. Countries must also develop their own initiatives—or delivery plans—to frame their actions and track the efficacy of their approaches.

Most importantly, countries and regions must call forward the leadership necessary to get the job done. They must measure not only the lagging indicators of progress, but also the leading actions that drive them to achieving the big goals. They must draw trajectories of progress and benchmark interim targets. They must convene their leaders and partners together in short, regular meetings to search for better and more effective ways to collaborate in light of the latest emerging truth.

In short, every country has to figure out what works. And all these actions can now be mapped, measured, and modeled thanks to GIS—new technology available to all.

Leadership from Nations Large and Small

The UNSD is currently teaming up with Esri and a pilot group of nations to define the requirements for this Federated Information System for the SDGs.

Initial implementation exercises will test a data hub that allows member states to measure, monitor, and report their progress on achieving the SDGs within the borders of their own geographic contexts—their own place, their own region, their own country.

The exercise also seeks to have national statistics offices align their data and systems with other in-country SDG stakeholders, including national mapping agencies, health ministries, natural resource and environmental agencies, and private-sector statistical data producers.

The common platform of location will allow member states and their various departments and agencies working on the SDGs to integrate data from different systems. The common denominator is location.

Sharing data-driven maps, apps, templates, and analytical tools. Applying maps and spatial reasoning to explore data in intuitive ways so they can gain a deeper understanding of the complexities and dependencies inherent in the SDGs. Parsing problems at different scales to test and discover which actions have the greatest impact.

This great work is an iterative and collaborative process.

One of the leading nations in this pilot group is Ireland—a country that has experienced its fair share of suffering and learning on the road to sustainable development. Other leaders in the pilot group include Brazil, Colombia, Kenya, Mexico, Morocco, Palestine, the Philippines, South Africa, Senegal, Tanzania, and the United Kingdom. Together, the pilot group is defining the requirements for an SDG Hub that will take on global dimensions—a system of systems.

The technology exists. The technology is not intimidated by scale. We must now land our data and our actions on the common platform—the map of our part of this earth. And we must quickly take those actions to scale.

Several members of the pilot group already have their SDG Hubs up and running. They are moving forward to build-out capabilities and tools to engage stakeholders from the public and private sectors, both domestically and internationally.

And there isn't a moment to lose.

Learn & Explore

George Washington on Educating Future Citizens and Public Servants
This article reveals how Enlightenment-based ideas on education for public administration are still quite relevant in thinking about education for public service today.

Sustainable Development Solutions Network
The UN Sustainable Development Solutions Network mobilizes global scientific and technological expertise to promote practical solutions for sustainable development, including the implementation of the SDGs and the Paris Climate Agreement.

A Data Hub for the Sustainable Development Goals
The United Nations Statistics Division has created a Federated Information System for the Sustainable Development Goals, also known as the SDG Hub.

For links to these and other examples, exercises, and resources, visit SmarterGovernment.com and click chapter 13.

The UNSD is working with a pilot group of nation-states—including Ireland—to quickly make available a Federated Information System for mapping and measuring performance toward achieving the Sustainable Development Goals.

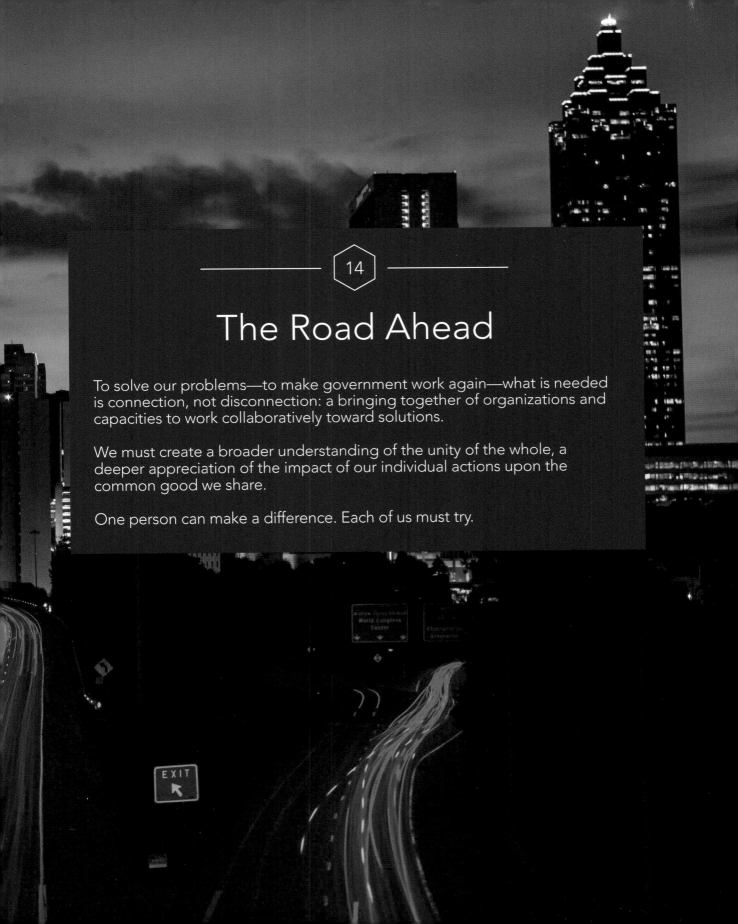

14

The Road Ahead

To solve our problems—to make government work again—what is needed is connection, not disconnection: a bringing together of organizations and capacities to work collaboratively toward solutions.

We must create a broader understanding of the unity of the whole, a deeper appreciation of the impact of our individual actions upon the common good we share.

One person can make a difference. Each of us must try.

A Blueprint for Smarter Government

We live in a time of profound change. A Third Industrial Revolution. The Internet of Things (IoT). The global rise of cities. The global rise of temperatures.

All around the world, urbanization and climate action are now joining together in one urgent movement of human development—the search for a more generationally sustainable way of life.

The challenges facing our world are many. And these problems will not solve themselves. But they can be solved. It is up to us to face them, to solve them, to respond to the changes of our times—the adversity of our times—with greater creativity, compassion, understanding, and action.

We live at a critical moment in this human story. No generation has ever had better technologies at their fingertips. No generation has ever before been so closely connected in a worldwide web of intelligence.

The ability not only to govern ourselves, but to govern ourselves well, is essential to writing the next and more life-giving chapter.

Creating Connections

In the not too distant past, many governors earned their reputations as reformers, in part, by breaking government into ever-more specialized parts. In the wake of those years of specialization as response to complexity, many legislatures created new executive offices to bring about a more coordinated and collaborative approach to problem-solving—an approach that specialization had inadvertently made more difficult.

For example, when departments whose mission it is to serve children, youth, and families—in Maryland's case, the Department of Human Resources (social welfare and child protective services), the Department of Health and Mental Hygiene, the Department of Juvenile Services, and the Department of Education—failed to effectively communicate, coordinate, and cooperate, the response was to create a Governor's Office for… Children, Youth, and Families.

When the various departments of public safety failed to work together, the response was to create a Governor's Office of Crime Control and Prevention.

And so on, and so on.

Creating coordinating offices can make us feel better, but they don't necessarily make things work better.

Of course, having sufficient numbers of executive staff is important. But putting separate, smaller bureaucracies on top of separate, larger bureaucracies doesn't necessarily guarantee greater effectiveness. In fact, creating more offices might just create more cracks in the bureaucracy for kids and other vulnerable people to fall through.

To solve our problems—and to make government work again—what is needed is connection, not disconnection: a bringing together of organizations and capacities to work more collaboratively towards solutions. This need is not only true for cities and states. It is true for nations and the great work of repairing our world.

Common platforms for collaborative action.

A system of connected systems.

An intelligent enterprise.

How Do We Measure Progress?

"Even if we act to erase material poverty, there is another greater task, it is to confront the poverty of satisfaction—purpose and dignity—that afflicts us all.

"Too much and for too long, we seemed to have surrendered personal excellence and community values in the mere accumulation of material things. Our gross national product, now, is over $800 billion a year, but that gross national product—if we judge the United States of America by that—that gross national product counts air pollution and cigarette advertising, and ambulances to clear our highways of carnage.

"It counts special locks for our doors and the jails for the people who break them. It counts the destruction of the redwood and the loss of our natural wonder in chaotic sprawl.

"It counts napalm and counts nuclear warheads and armored cars for the police to fight the riots in our cities. It counts Whitman's rifle and Speck's knife, and the television programs which glorify violence in order to sell toys to our children.

"Yet the gross national product does not allow for the health of our children, the quality of their education or the joy of their play. It does not include the beauty of our poetry or the strength of our marriages, the intelligence of our public debate, or the integrity of our public officials.

"It measures neither our wit nor our courage, neither our wisdom nor our learning, neither our compassion nor our devotion to our country; it measures everything in short, except that which makes life worthwhile.

"And it can tell us everything about America except why we are proud that we are Americans.

"If this is true here at home, so it is true elsewhere in world."

—Robert F. Kennedy, University of Kansas, March 18, 1968

Measuring Genuine Progress

The philosopher Pierre Teilhard de Chardin, S.J., once said, "And yet it is the law of all progress that it is made by passing through some stages of instability."

In many ways, we post-industrial people like to think of ourselves as pro-growth. We believe in growing jobs and growing opportunity. We believe in children growing healthy, growing educated, and growing strong. We believe in grandparents growing old with dignity, security, and love. Increasingly of late, we believe in growing more trees, growing sustainable fisheries, growing food locally to feed our citizens. And, as Americans, we still believe in growing prosperity for every generation.

How do we measure this sort of growth? How do we distinguish between increased consumption and genuine progress—the sort of progress that makes life better for our children and grandchildren?

Gross domestic product (GDP) is an important measure, but it cannot be the only measure. GDP tells us what we are producing, but it neglects the cost of what we are using up—clean air, clean water, natural lands and forests, the health of our people. GDP tells us the monetary value of what we are extracting from the earth, but it doesn't tell us whether we are creating greater opportunity for all, or greater security, health, and well-being for this generation and the next.

Our country's GDP has doubled over the last three decades. Yet things like income inequality, wages, middle-class opportunity, the amount of carbon pumped into our atmosphere, and life expectancy are not moving in the right direction.

Jeremy Rifkin, renowned economic planner and author of *The Third Industrial Revolution*, goes so far as to say that GDP is a better measure of our indebtedness to the earth than it is of generational health, prosperity, and well-being.

There is a difference between consumption and well-being. There is a difference between income—which is fleeting—and wealth, which is lasting. States and nations do not build wealth by storing cash in a bank vault. They build wealth by expanding human solutions to human problems so more and more of a nation's people—living and yet to be born—are able to enjoy a higher quality of life.

In 2010, Maryland became the first state in the union to develop and adopt a Genuine Progress Indicator (GPI). Many other states have now done the same. In Maryland, we began to measure not only job and wage growth, but also whether the health of our air and water was increasing. Whether traffic congestion was growing worse or mobility was getting better. Whether Maryland was maintaining and growing its forest cover and tree canopy.

Genuine progress, like the Star-Spangled Banner, consists of many strands, many stripes, and a constellation of stars—all part of one whole. Our GPI became for us a more holistic tool for measuring progress than GDP; a better guide for policy and budget choices; a better way of evaluating the health of the common good we share.

A system without feedback loops eventually fails. This is why we began to integrate GPI into our decision-making process for budget and policy matters alongside the simpler exercise of balancing the budget.

Our country, our states, our cities are all systems. A GPI gives us the feedback loops that tell us whether or not we are succeeding—provided leaders seek, and allow, an honest assessment of whether our graphs are moving in the right or wrong direction.

By setting operational goals and measuring leading actions on a daily basis, we were able to improve the effectiveness of our state government. But a GPI measures much more than that. It measures the quality of life we share as a people. It allows us to measure societal progress, environmental progress, and the expansion of human solutions to human problems. It allows us to answer questions that

What Is GPI?

The Genuine Progress Indicator (GPI) is a metric that can be used to give a more accurate picture of the health, well-being, and progress of a community or society than by looking at gross domestic product or GDP. GPI takes into account environmental factors, such as pollution, resource depletion, carbon footprints, and sustainability, as well as personal economic factors—inputs ignored when calculating GDP, which simply focuses on the size of the economy.

Some of the factors that go into calculating GPI include:

- Personal consumption
- Crime
- Air, water, and noise pollution
- Environmental damage
- Poverty
- Income distribution
- Housework, volunteering, and higher education
- Loss of farmland and wetlands
- Changes in leisure time
- Lifespan of consumer durables and public infrastructure

Economic Indicators	Environmental Indicators	Social Indicators
Personal Consumption Expenditures	Cost of Water Pollution	Value of Housework
Income Inequality	Cost of Air Pollution	Cost of Family Changes
Adjusted Personal Consumption	Cost of Noise Pollution	Cost of Crime
Services of Consumer Durables	Cost of Net Wetlands Change	Value of Volunteer Work
Cost of Consumer Durables	Cost of Net Farmland Change	Cost of Lost Leisure Time
Cost of Underemployment	Cost of Net Forest Cover Change	Value of Higher Education
Net Capital Investment	Cost of Climate Change	Services of Highways & Streets
	Cost of Ozone Depletion	Cost of Commuting
	Cost of Non-renewable Energy Resource Depletion	Cost of Motor Vehicle Crashes
		Cost of Personal Pollution Abatement

There is a more holistic measure of progress beyond gross domestic product or gross state product. In 2010, Maryland became the first state to adopt the Genuine Progress Index (GPI)—a measure of 26 economic, environmental, and social indicators.

humanity has wrestled with since the beginning of recorded time—questions like "Where are we going?" and "How do we know we're getting there?"

In our state, we measured the value of higher education, the cost of crime, income inequality, the cost of ozone depletion, the value of volunteer work, and score of other indicators—from clean air to healthier waters. There were, in all, twenty-six quality-of-life measures spanning economic, social, and environmental progress.

In city after city, mayors like Joe Curtatone of Somerville, Massachusetts, and others are instituting a citizen happiness index to gauge public satisfaction with city living.

Biologist and author Janine Benyus once said, "Life creates conditions conducive to life." Today, we have within our power—as cities, as states, as communities, as individuals—the ability to achieve rising standards of living, better-educated children, more affordable college, a more highly skilled American workforce, safer neighborhoods, a safer and more resilient homeland, healthier people, and a more sustainable way of life with the other living systems of the earth.

But this will not happen on its own.

Leadership is important. Broader understanding is essential. Progress is a choice. And better choices are possible when they are guided by more holistic measures of the common good we share.

Measuring Genuine Progress in Maryland, 1960 to 2013

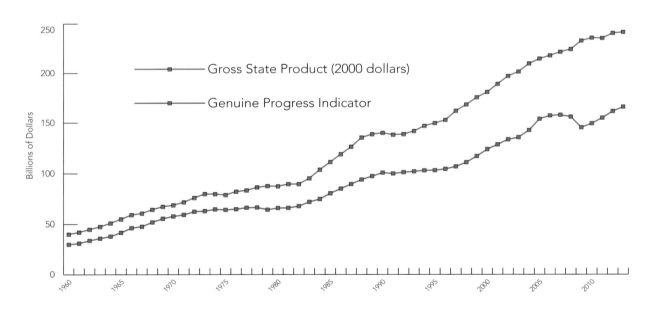

Maryland's gross state product has grown from less than $30 billion in 1960 to more than $165 billion in 2013. But although the Genuine Progress Indicator has increased significantly over the past 50 years, it has failed to keep pace with the increase in the gross state product of Maryland. This is attributable to increases in income inequality, the overall loss of natural lands, the costs of climate change, and the depletion of nonrenewable natural resources.

The Evolution Continues

Adversity is the greatest teacher.

One of the sharpest commanders in American policing today is Deputy Chief Sean Malinowski of the Los Angeles Police Department (LAPD). Deputy Chief Malinowski, like the great Jack Maple, has a gift and intuition for understanding how to move large human organizations forward. He understands how to lift up the leaders. How to set a cadence of accountability. How to see the connections between the connections. He also understands how technology can help his officers get inside the turning radius of dynamic problems—in his case, crime patterns.

He is a veteran of more than two decades who has seen the ups and downs of policing and police/community relations. He has also seen something else—he has seen his department evolve after its eyes were opened to a new way of policing by LAPD Chief Bill Bratton, who brought CompStat to Los Angeles.

Over the last couple of decades, the LAPD has never stopped evolving, never stopped adapting, never stopped pushing for better and more effective ways—enabled by technology and the practical science of what works.

In one concise illustration, Malinowski describes the LAPD's operational evolution since 2012 like this:
> "We have moved," he says, "from making decisions based on hunches to making decisions based on evidence. From information that was guarded and locked in separate silos, to an organization where information is shared by all. We have gone from simply answering 911 calls for service to deploying—more rapidly and proactively—based on alerts. And we have moved from making pretty maps of where crime happened yesterday to using predictive analytics to tell us where to prevent crime on the map before it happens tomorrow."

Deputy Chief Malinowski's slide isn't just about Los Angeles or the LAPD. His single slide summarizes the evolution of smarter governing at every level of government in the Information Age.

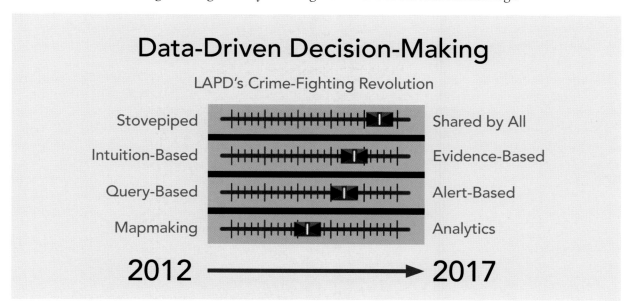

This illustration is borrowed from Deputy Chief Sean Malinowski of the LAPD. Although it describes the recent organizational evolution of the LAPD, it also describes the evolution of a new way of governing in well-led organizations around the world.

Sensors and Alerts

Over the last two years, I've had the honor of working with more than forty leading cities and their university partners in a smart cities collaborative called the MetroLab Network. This coming together grew out of an Obama White House initiative—specifically a report by the President's Council of Advisors on Science and Technology. Their vision was to engage a collaborative web of cities and universities to share leading practices, learn from the experiences of one another, and accelerate progress.

There is a common international vocabulary and mission statement emerging to define what is meant by the term "smart city." Smart cities are those that use modern technologies—like big data, GIS, and the IoT—to make their cities more inclusive, more prosperous, more resilient, and more sustainable. Smart cities strive to make their people healthier, more secure, more educated, more highly skilled, and more mobile.

If city/university partnerships 101 was mostly about urban real estate redevelopment, MetroLab believes city/university partnerships 201 will be all about smart city solutions to big city challenges—using new technologies to solve problems. Specifically, directing the firepower of university research and development talent to tackling the mayor's biggest operational, social, and environmental challenges; using big data and big data analytics—not just to study big city problems, but to solve them. And all of this is enabled by GIS.

Enter the wild world of sensors and alerts—our newfound ability to monitor, in real time, the dynamics of a city through the deployment of electronic sensors. What kind of sensors?

Sensors—powered by small solar panels—that alert city cleaning crews when a public trash can needs emptying. Audio sensors that can instantaneously pinpoint—with accuracy and certainty—the exact location where a gunshot has been fired. Sensors that monitor the value of green infrastructure—like porous alleyways and rain gardens—for the absorption of stormwater that would otherwise poison streams. Sensors that monitor traffic flows to reduce congestion and turn traffic control lights into dynamic systems that optimize the flow of cars. Sensors that measure air pollution and its concentrations across a city.

Motion sensors like the ones deployed in Singapore that protect the well-being of senior citizens living alone. Sensors in ankle bracelets that are used to make sure vulnerable children on juvenile probation are obeying their curfews and not hanging out on dangerous corners where gunfire flies.

One big-brain engineering professor at Carnegie Mellon University showed me the low-cost and highly accurate sensors that the university and the city are deploying across the many bridges that span the three rivers of Pittsburgh. "These sensors," she explained, "allow us to feel when a bridge is experiencing pain..."

Another team of engineers is working on the very real technology of sensors that ride on the city's fleet of vehicles and relay a nonstop flow of data on the conditions of the roads and bridges over which they travel.

Doing something about what we know is still the key, but sensors are giving us a whole new way to know—and at a scale and speed never before possible.

Of course, iPhones® in the hands of caring citizens are a pretty powerful feedback loop, as well. But automated sensors are a new and more cost-effective way to better manage moving things. Predictive analytics is another.

Predictive Analytics

Shortly after the attacks of September 11, I had occasion as mayor to sit down for a frank discussion about information sharing (or more exactly, the lack of it) between local, state, and federal law enforcement. One of the career federal intelligence officials at the table said, "If we only knew what we already know..."

Predictive analytics is the ability to understand what we already know. It is the science of probabilities—based on past experience—applied to present and future events.

Former mayor of Indianapolis Steve Goldsmith, now a professor at Harvard's John F. Kennedy School of Government, keeps a running compendium of predictive analytics use cases from cities across our country. Predicting where structural fires are most likely to happen, predicting where food-borne illnesses are most likely to strike, predicting which neighborhoods are most likely to flood, or where abandoned and vacant housing is most likely to cluster.

With predictive analytics, The Science of Where® meets the science of why—when we understand what happened where, why, and... *when* before.

Let me show you what I mean.

The Birth of PredPol

By 2008, forward-looking police leaders began to ask the question: Can we predict when and where crime will happen in the future, based on where and when it has happened before? Bill Bratton asked a couple of university professors if they could more precisely pinpoint where crime was likely to happen over any part of Los Angeles over any given patrol shift. The mathematicians said "yes, if"—if you can give us ten years of data on where and when crime happened before.

Of course, with 911 and a record of geo-located calls for service, the data existed, the data was public,

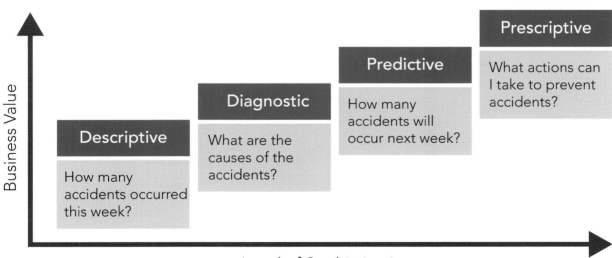

What do we mean by "predictive analytics"? This model from the management consulting firm Grant Thornton, LLP, lays out the evolutionary progression in the sophistication and the value of predictive analytics. In this case, the goal is figuring out how to reduce traffic accidents.

and the data was provided.

When the mathematicians came to understand what the LAPD data already knew, they saw that the commission of crime actually followed predictable patterns of time and space. In other words, there was something about the built environment (place) and the time of day (opportunity) that followed a predictable pattern of crime commission—even with the huge variables of criminal intent and human will.

They developed an algorithm—a fancy word for "mathematical formula of probability"—that allowed the LAPD to create not merely a generalized heat map of where crime *had happened* over the last ten years, but a predictive map of where crime was most likely to happen in tiny 500-foot-by-500-foot squares, over the next eight hours, for this day of the year, over any given patrol district in the city.

Soon, the new technology was deployed throughout the LAPD. Commanders could see the "red boxes" in every patrol district. Police officers would be given the maps at roll call and be asked to spend eighty-one quality minutes—windows down, eyes open—patrolling inside the boxes whenever they were not responding to 911 calls for service.

The maps were ground-tested with the officers. They worked. A new company called PredPol was born, and more and more cities are now deploying the technology.

Protecting Children in Pittsburgh and Auckland

Second example: What do Allegheny County (Pittsburgh), Pennsylvania, and Auckland, New Zealand, have in common? Turns out, it is a concern for vulnerable kids who die or who are abused when their cases are not properly assessed when first referred to child welfare.

Together, Allegheny County developed a risk assessment tool with the help of researchers at the University of Auckland that would give proper weight to a variety of factors in a child's and family's history. This tool assures that better decisions are made about the level of investigation and intervention required to protect a child.

Their predictive formula, based on objective criteria, is open source—anybody, anywhere can plug their own county or city's experience with kids into the equation, and see if they can't do a better job of saving vulnerable young lives.

Allegheny County, with Auckland's help, also controls for racial bias. In other words, they track race and ethnic background not as an element of the predictive equation, but to make sure the predictive equation does not effectuate a racial bias.

Preparing for Earthquakes in New Zealand

My third example—speaking of New Zealand—is the Kaikoura Earthquake, which rocked Wellington, New Zealand, in 2016.

Sensors, building data, soil depth, and geological depth between the ground and the bedrock—along with the cruel ground test of an actual earthquake—have all combined to give the Wellington City Council a tool for predicting proper building standards across the development footprint of their entire city.

Predictive analytics now tell Wellington which places will shake and quake the most—places where no buildings should ever be built; which places will shake and quake the second most—where much stronger building codes need to be put into place; and those neighborhoods that will shake and quake the least—where existing quake-proof building codes are sufficient.

Predictive Analytics in Action

The Kaikoura Earthquake near Wellington, New Zealand

This map (left) shows the depth to bedrock and how the shape of that bedrock affects damage on the surface, to better inform planning, preparation, and resiliency.

This map (right) that shows terrain angles, along with major slips, is a vital new tool for better preparing for and managing building decisions in anticipation of the next quake.

3D data and GIS tools are now used to better visualize and understand the predictable impact of future earthquakes on buildings and their surrounding areas (left).

A Catalyst for Smarter Communities

No one cares about a neighborhood as much as the people who live in it. Our communities are where we spend our daily lives—going to work and school, walking our dogs, raising our children, making what often amounts to the largest financial decision of our lives, and sleeping every evening. Neighborhoods are part of an urban tapestry, working together to build diverse, resilient, and livable regions.

ArcGIS Hub initiatives are focused, policy-driven goals that stem from executive and strategic objectives.

Residents and citizens want to actively engage with government to understand policies, share their local knowledge, and improve their communities. There is a tremendous opportunity to use modern technology, combined with the collective knowledge of governments, citizens, and researchers, to build smarter communities.

How do smarter communities around the world encourage effective governments to work closely with residents, and vice versa? All facets of policy making, people, processes, and technology need to be in balance and aligned to make engagement efforts honest, supported, and successful. So while innovative technology can dramatically improve citizens' access to government and the overall success of such partnerships, it needs to be deployed in a way that supports shared outcomes.

It Starts with a Goal

Open data is a global movement to share authoritative information that is public and reusable to drive economic development, improve government efficiency, and increase stakeholder engagement. By itself, open data is a great set of principles, but the concept often lacks focused goals or expected outcomes. Open data is much more effective when it is driven by strategic priorities and empowered to measure outcomes for improving engagement on specific, constituent-centered benefits.

ArcGIS Hub is a new, two-way engagement platform that truly connects government and citizens. It introduces a new framework designed to prioritize initiatives, organize data and teams, measure the progress of key performance indicators, and empower the community to understand complex relationships with explorative analytics and infographic reports. It includes tools for managing open data, but it goes beyond just open data. It provides a catalyst for creating smarter communities.

Initiative-Driven Engagement

ArcGIS Hub initiatives are focused, policy-driven goals that come straight out of executive and strategic objectives. Vision Zero, for example, is a global initiative to prevent traffic-related deaths and life-altering injuries. The goal—to have zero traffic-related deaths—is clearly identifiable as an important and measurable metric. Everyone has a vested interest in accomplishing this target, and as a result, groups are motivated and focused on how they can work together to achieve this outcome.

In addition to aiming data and apps at a specific goal, ArcGIS Hub initiatives can help governments better coordinate their own work and invite other organizations, such as civic advocacy groups, to cohost events, gather new data, perform analysis, and collaborate with tech developers to devise creative solutions to issues. Discussion and feedback from the community are captured through ArcGIS Hub conversations, which are forums to track public feedback, government responses, and ArcGIS Hub initiative team actions to ensure that everyone is working together toward the initiative's goal.

More Cases in the United States

A few final examples.

The New York Police Department now does a simple short-form "public opinion poll"—riding on free iPhone apps—that tells commanders on a rolling basis the police precincts where public trust in the professionalism and courtesy of neighborhood police officers is trending up, and where it is trending down. A shaded map now informs other flows of data and knowledge like trends in citizen complaints and ongoing internal investigations. If public trust in police is tanking, and commanders don't know why, they now know to dig deeper.

In Cincinnati, Ohio—and many other places, actually—hospital, Medicaid, and social services data is being combined to target prenatal and infant care support to moms whose babies are in greatest danger of dying before their first birthdays. The reductions in infant mortality rates have been impressive everywhere these sorts of predictive analytics and strategies have been followed.

In Philadelphia, Pennsylvania, researchers have used data on home sales, property values, and rental rates to predict where gentrification is most likely to happen at the fastest pace. The ability to predict rising values might well inform inclusionary zoning and other affordable housing decisions to be deployed in more targeted and effective ways.

In Maryland, the Chesapeake Conservation Center is using years of Landsat photography and modern lidar images to figure out the natural hydrology—flow of water—across any tract of land. This newly discovered ability to map the way rainwater has "always" flowed off of a tract of agricultural land, for example, allows farmers to pinpoint exactly where to install sediment ponds or to restore wetlands to prevent nitrogen, phosphorus, or sedimentary pollution.

Predictive analytics is the tool; precision conservation is the action.

Watching Out for Secret Formulas

A cautionary word on predictive analytics: beware of secret algorithms.

No mathematician will ever enjoy so much credibility and trust in the public arena that he or she can escape the new standards of openness and transparency. If your probabilistic formula has to remain secret, it probably isn't good enough for public-sector deployment in the Information Age.

Openness and transparency isn't just a proverb. It's an indispensable requirement for public trust in public administration—especially when deploying new methods and new technologies. The only respected authority today is "I can show you it works."

Inviting Artificial Intelligence to Stat Meetings

One more technological advancement for governing our cities, counties, states, and countries deserves mention. It is the emergence of artificial intelligence, or AI. AI uses advanced computing techniques to enable computers to discern and understand patterns in ways that people gathered around a CitiStat or CompStat meeting might. But the computer is potentially able to do it faster, and with a deeper memory.

How might AI be integrated into future stat systems? Let's take a look at an example from Marc Benioff, CEO of Salesforce. He holds weekly meetings with executives and senior staff to review goals and measure progress. One seat at that conference table is reserved for Salesforce's AI software, named Einstein. Einstein has access to all the company's data including sales numbers, and is frequently called on not only to verify information presented by other participants in the meeting, but also to provide deeper insights.

It's hard for me to imagine a similar AI bot sitting at the table during a reinstituted CitiStat meeting in Baltimore. I suppose we would probably start it out in the CitiStat projection booth where other young CitiStat analysts begin. They could whisper insights into the meeting by way of notes or texts as the conversations unfold.

Then again, maybe—after it passes its internship—we'd dress it in a homburg hat and spats, sit it at the table, program in a surly Brooklyn accent, and call it "Jack"—after the venerable Jack Maple, who created CompStat in New York back in 1994. (Full disclosure, Jack actually imagined and gave voice to this future scenario... after a few drinks. He also imagined a performance-measurement musical that would one day wow Broadway.)

Sadly, in today's reality, when department heads quit or get fired, or administrations change, there's a lot of "institutional memory" that gets wiped away. Maybe AI stat-bots could save us from that inevitable brain drain? Not only would "Jack the AI stat-bot" have access to data about the performance metrics and customer service requests of the current administration, but it would also have a deep library of knowledge that goes back into time—well beyond the collective memory of any current assortment of administrators and department leaders.

"Jack the AI stat-bot" would know which work crews had their highest performance levels, when, and under which leaders. It would know what has been done in the past, what worked, and what didn't work. The internet would allow it to network with other AI stat-bots (I'm thinking "Lou Anemone the AI stat-bot," "Bill Bratton the AI stat-bot," or "Matt Gallagher the AI stat-bot"—he'd be a killer.) These stat-bots might be in the employment of other cities or states across the country—all of them working on real-life problems in real time on real maps.

"Jack the AI stat-bot" might access the knowledge and experience of other cities that have successfully tackled homelessness, or drug addiction, or lead poisoning. It might help everybody on the bridge of the fictional Starship *Enterprise* of *Star Trek* fame make better decisions based on a much larger pool of real-world experiences and better data.

Is it science fiction or just a view over the horizon? It's not such a big leap forward from Alexa to "Jack the AI stat-bot."

Show Me My House

So where does all this take us? Probably back to where we started—to "know the place for the first time," as T.S. Eliot once said.

Back in the earliest weeks and months of CitiStat, we used to hold regular community meetings inside City Hall that we dubbed "Mayor's Nights In." Unlike "Mayor's Nights Out," which were open to the general public and held in high school gymnasiums throughout the various neighborhoods of town, "Mayor's Night In" was by special invitation.

It was about lifting up the neighborhood leaders. On these nights, we invited the neighborhood leaders of the council districts into City Hall on a rotating basis—community association presidents, vice presidents, treasurers, secretaries. And of course, we would intentionally hold these meetings in the CitiStat Room so that we could show these leaders Baltimore's new way of governing and getting things done.

On each of these nights—with leaders drawn from every district of the city—there was always one question. Whenever we would start to show our neighbors the smart maps, the graphs, and the aerial photographs arrayed on the big screens of the CitiStat Room, this one question would come up without fail. And it would come early on.

Regardless of whether a neighborhood was black or white, or rich or poor, or Democratic or Republican, someone—within the first five to ten minutes of my map-laden presentation—would always raise their hand to ask, "Can you show me my house?"

Good question. The essential question. The opening and closing question—really—for a government of, by, and for the people.

Many times, that question was voiced less in the form of a question and more in the form of a demand: "Show me my house."

Why was that?

Was it to know that my government works?

Was it to know that I matter to those whom I elect to serve me by running my government?

Was it to better understand how I am connected to the people and events around my neighborhood and city?

The freedom of the Exodus story was not merely a freedom from oppression; it was a freedom for full participation in the shared life of a community.

"Show me my house."

Perhaps it is the demand at the heart of one of the most important beliefs we share as a people—

Whenever we brought groups of neighborhood leaders into City Hall to provide a demonstration of the map and the method of CitiStat, someone in the group would always raise their hand to ask the ground-truthing and essential question: "Can you show me my house?"

the belief we share in the dignity of every individual person. The belief that one person can make a difference, that each of us is needed. And "we must help each other if we are to succeed."

Technology—whether GIS or the internet—is not the end in itself. It is, however, the means of a new beginning. A new and better way of governing and getting things done. A crowd-sourced healing of the deepest kind.

"Show me my house."

Start, and don't stop.

Lift up the leaders.

And lead.

Learn & Explore

MetroLab Network

The MetroLab Network helps drive partnerships between local governments and universities. MetroLab is poised to help the public sector adapt to rapid technology change.

A Platform for Community Engagement

ArcGIS Hub is a two-way engagement platform that helps connect connect government and citizens.

For links to these and other examples, exercises, and resources, visit SmarterGovernment.com and click chapter 14.

The City of Brampton in Ontario, Canada, lets citizens share their ideas and vision for the future by highlighting what they love, what they wish they had, and what they think the city could do better.

Appendix A: Study and Discussion Questions

Chapter 1: A New Way of Governing
- What are the four tenets of any performance management or "stat" process?
- What does a "cadence of accountability" mean?
- What are the five hallmarks of a collaborative culture for progress?
- What is a "policy map"?
- What are the benefits of open data portals?
- What is GIS?
- What is the problem-solving technique known as the 5 Whys?
- How has the Information Age changed the nature of effective leadership?
- What were the characteristics of the old way of governing and getting things done?
- What are the characteristics of the new way of governing and getting things done in the Information Age?
- How does the new way of governing compare or contrast to the old way of governing?
- What is the most important challenge facing your neighborhood that the new way of governing could help address?
- What are the two technological developments that The Science of Where® has brought together for policymakers and executives?
- How does the combination of these technologies allow for better collaborations, at scale, in real time?
- Describe the evolution underway—in better-led organizations—thanks to the increased use of data, the map, and the method of performance management?

Chapter 2: When Disaster Strikes
- In an emergency, what is the first question?
- What role did the map play in the aftermath of the Baltimore Howard Street Tunnel Fire?
- What common challenges make coordination and collaboration difficult when it comes to solving dynamic and complex problems during an emergency?
- What is the common operating platform for bringing together separate silos of information?
- What did OSPREY allow the State of Maryland to do?
- How is OSPREY an example of the new way of governing?
- What is the "all crimes, all hazards" approach?
- What were the goals or core capacities of Maryland's drive for improving homeland security and preparedness? Were there any capacities that surprised you? Were there core capacities that should have been included? Why?
- In a crisis, how would you decide what information and actions to prioritize?
- Given the rapidly changing social, political, and technological landscape of the world, how would you update emergency preparedness to match evolving threats?

Chapter 3: Collaborative Leadership

- What is your personal definition of leadership?
- What are the two essential questions for all leaders today?
- What are the "eight drivers of leadership"? Define each of these qualities or disciplines; use examples from your own experience.
- Why is solitude essential for creative and impactful leadership?
- What are the three new rules for leadership?
- What is meant by leadership "presence" and why is it so important?
- What is the difference between a dream and a goal?
- Have you run into a culture of excuses in your workplace? How have you overcome it?
- How does a good leader foster a collaborative culture for progress?
- What are the human failings that kill a collaborative culture?
- How does a good leader safeguard against these vices?
- What are the essential powers a good leader must exercise at the center of the collaborative circle?
- How do the "eight drivers of leadership" factor into these powers and practices?
- Are good leaders born or are they made?
- What are the virtues and disciplines of effective leadership in every age?
- What criteria would you use to assess the effectiveness of a team?
- How is the setting of a "cadence of accountability" an exercise of leadership power and presence?
- Describe how being open to questions can fuel the drive for progress.
- Can you think of any instances in your work where open and regular questioning led to revelations that ultimately improved performance?

Chapter 4: A New Way of Policing

- Who was Jack Maple?
- What is CompStat?
- What were the newly emerging technologies that allowed CompStat to go real-time and to scale?
- What is the common platform of technology used by CompStat?
- How did CompStat change the nature of police deployment and crime-solving?
- How did CompStat use data, mapping, and deadlines to establish a tempo of operational progress?
- Why is the method of CompStat described as both entrepreneurial and agile?
- Describe some of the ways CompStat changed policing for the better.
- What were some of Jack Maple's views on leadership?
- What are the four tenets of any "stat" process?
- How would you ensure data integrity in a system that rewards improving metrics?
- What criteria would you use to assess the effectiveness of CompStat?

Chapter 5: Making Baltimore Safer
- What are the key elements of a CompStat Room?
- What is a "whole of government" approach?
- What was "Chalkman," and how did it influence Baltimore policing and city government?
- What is Jack Maple's "canon for crime-free communities"?
- What were some of the leading actions that Baltimore put into place to measure and improve the integrity, professionalism, and courtesy of its police force?
- When were these measures discontinued and why?
- What are the leading actions every city/county should put into place to improve police professionalism, efficacy, and public trust?
- What actions can our governments take to improve public trust and reduce use of lethal force?

Chapter 6: CitiStat and the Enterprise of Governing
- What was the first city in America to apply the tenets of CompStat to the entire enterprise of city government?
- What is CitiStat?
- What were the elements that made CitiStat an innovation in government?
- In what ways did the creation of a 311 call center inform the CitiStat process and build public trust/confidence in local government?
- What are the steps for implementing a CitiStat system?
- What are the characteristics of a robust CitiStat process?
- What are the biggest cultural, political, and technological barriers to implementation?
- Why is it critically important to lift up and highlight the work of leaders within the organization?
- How does CitiStat bring visibility and recognition to the highest performers?
- What is the "wheel of accountability"?
- How could a stat process be applied to improve your organization?

Chapter 7: Taking It Statewide
- What is StateStat?
- Which job did Governor O'Malley enjoy most—mayor or governor—and why?
- Why was it important to declare and own a limited number of strategic goals for the state?
- Why was it important to maintain public facing dashboards on each of the sixteen strategic goals for the state?
- How can citizen awareness of progress to goal contribute to public support?
- How can stakeholder awareness of leading actions and leading measures contribute to better policy, increased investment, and more effective regulations?
- Define the term "chain of delivery."
- What is a "delivery plan"?
- How does StateStat compare or contrast to CitiStat? How were they similar and how were they different?
- What is the tension between "the whirlwind" and the organizational focus required for making progress?

Chapter 8: Improving Education

- Write a brief outline of the framework for educational progress in Maryland.
- What were the "lagging indicators" that Maryland pursued to improve college and career readiness by 25 percent by 2012?
- What were the "leading actions" Maryland schools took to make progress toward the lagging indicators?
- What role did Maryland's P-20 Leadership Council play in developing performance measures for progress?
- How did measures of "transitional readiness" play into the mission of continuous educational improvement?
- How does the framework for education compare or contrast to the frameworks for improvement for the other big goals in this book?
- What are the elements of a good "delivery process"?
- What is the difference between a "leading action" and a "lagging indicator"?
- What is the difference between a "delivery goal" and a "key performance metric"?
- What evidence is there that things got better in Maryland for kids?
- How would you design a new system for measuring educational progress?
- Do you agree with the measures of success outlined in this chapter?
- How would you determine whether an education system is improving or not?

Chapter 9: Improving Health and Well-Being

- What are the key questions to ask about any public health challenge?
- How does the framework for improving health and well-being compare or contrast to the frameworks for improvement for the other big goals in this book?
- What evidence is there that Maryland improved health and well-being?
- How would you design a system for measuring health and well-being?
- Do you agree with the measures of success outlined in this chapter?
- How would you determine whether health and well-being are improving in your community?
- What was the fundamental shift in the way Maryland began to reimburse hospitals for patient care?
- What were the building blocks and events that allowed leaders in Maryland to make this shift?
- What might this shift portend—make possible—for the use of predictive analytics to improve patient outcomes and well-being?

Chapter 10: Restoring Our Waters

- What are the three primary types of pollution that degrade the health of the Chesapeake Bay?
- What are the four main sectors of our economy/landscape that contribute this daily load of pollution?
- How does the percentage of contribution from each source vary from one rivershed to another?
- What is the "dead zone" and how is it created?
- What were the major changes Maryland brought to the effort to restore the Bay from 2007 to 2014?
- How did this approach differ from past efforts at Bay cleanup?
- What role did political leadership play in securing a multistate agreement to measure performance in short intervals with openness and transparency?
- What are the thirty-three leading actions that BayStat mapped, measured, and drove to improve Bay health?
- What are the six indicators used by environmental scientists to measure the health of the Bay and its rivers?
- What is BayStat?
- What are the "four disciplines of execution"?
- What are the similarities and differences between the four disciplines of execution and the four tenets of any "stat" process?

Chapter 11: Preserving Our Land

- What does "land use" refer to?
- What was Maryland GreenPrint?
- Why is preserving our land so important?
- What is green infrastructure?
- How can mapping be used to build public consensus for land preservation?
- What functional and political advantages can leaders harness by publishing maps online for all to see?
- Differentiate between smart growth and green infrastructure.
- What are the preservation challenges facing your community?
- What solutions would you suggest for the balance between growth and preservation?
- What criteria would you use to assess the success of land preservation?
- How does the work to preserve land in Maryland compare or contrast to the work on other big goals outlined in this book?

Chapter 12: Protecting Our Air

- What is the greatest business opportunity to come to the United States in one hundred years?
- Outline the steps Maryland took to protect our air.
- How would you design a new system to protect our air?
- What criteria would you use to assess progress in reducing pollution?
- What are the policies and actions that states can take—on their own or with other states—to improve air quality and reduce greenhouse gas emissions?
- What are the measures that would inform the public as to whether those actions are working over the long term to achieve the desired goals?
- What were the goals and the leading actions Maryland took where climate change, clean air, and the reduction of greenhouse gases are concerned?
- Is Maryland still making progress on these fronts?

Chapter 13: The American Revolution, and Our World

- In what ways did George Washington's leadership compare or contrast to the new way of governing outlined in this book?
- Outline the components of George Washington's propositions for effective public administration.
- In what ways are effective self-governance and a public administration a "longitudinal experiment"?
- Do you agree that public administration is a science? Why? Why not?
- Describe the relationship George Washington saw between government legitimacy and government effectiveness.
- How do George Washington's propositions of good governance compare to the Worldwide Governance Indicators espoused by the World Bank today?
- How does the early implementation of FedStat in 2015 compare or contrast to the full vision of what a stat program can achieve?
- What metrics do you believe would be most important to measure at the federal level in a FedStat program?
- What criteria would you use to assess the success of a FedStat-type program?
- What are the United Nations Sustainable Development Goals?
- Is it possible to achieve any of these goals with a performance management system?
- What role does political leadership play in achieving the Sustainable Development Goals in any country or city?

Chapter 14: The Road Ahead

- What is needed to make democratic government work again in the eyes of its citizens?
- Outline the steps that need to be taken to improve our government at the local, state, and federal level.
- What is the future of smart government?
- What ideas can you add to the road ahead?
- What are the ethical concerns, disparity concerns, and community concerns that public leaders must address when using predictive analytics?
- How can these concerns be addressed?
- Are there "dos and don'ts" of policy and practice that should be followed when it comes to using predictive analytics?
- What would a model policy for the implementation of predictive analytics look like?
- How can effective reforms and best practices be institutionalized in representative democracies so they survive electoral transitions?
- Is there any way to protect new systems from being disbanded as large organizations revert to "the way we've always done it"?

Appendix B: GIS Exercises

Chapter 1: A New Way of Governing

The Power of Maps
ArcGIS® comes with a set of ready-to-use apps that enable people to work with online maps for a variety of roles. ArcGIS also enables people across government and beyond to create their own apps for specific tasks and audiences. Try this lesson to explore how GIS maps work, and learn what you can do with different types of maps and apps.
https://learn.arcgis.com/en/projects/the-power-of-maps/

Chapter 2: When Disaster Strikes

Collect Hurricane Damage Assessments
Hurricane Harvey hit Texas in August 2017. The hurricane caused an estimated $125 billion in damage and destroyed 12,700 homes. In major disasters, multiple organizations work together to assess the damage. Collecting and sharing information efficiently is necessary for these organizations to make informed decisions. In this lesson, you'll use multiple ArcGIS apps to create, collect, and share hurricane damage assessments of residential buildings.
http://learn.arcgis.com/en/projects/collect-hurricane-damage-assessments/

Identify Landslide Risk Areas in Colorado
In this lesson, you'll learn to identify areas at highest risk of flooding and landslides during a major rain event in Boulder County, Colorado. First, you'll explore the region's dramatic geography and identify how the most flood-prone areas correspond with large population centers. Then, you'll determine where landslides are most likely to occur, and you'll summarize the population in these at-risk areas. Overall, you'll get a picture of the dangers that people face in Boulder County and the costs associated with rebuilding in the wake of catastrophe.
https://learn.arcgis.com/en/projects/identify-landslide-risk-areas-in-colorado/

Analyze Volcano Shelter Access in Hawaii
In this lesson, you'll explore the relationship between shelter locations and population in high-hazard lava flow zones near Mount Kilauea in Hawaii. In particular, you'll look for areas where the risk is high and shelter access is poor.
https://learn.arcgis.com/en/projects/analyze-volcano-shelter-access-in-hawaii/

Assess Open Space to Lower Flood Insurance Cost
In this lesson, you'll perform an analysis that will identify protected open areas in Georgetown County, South Carolina, that are eligible for Open Space Preservation credit. First, you'll connect to the ArcGIS® Living Atlas of the World in ArcGIS® Pro and modify the layer using a raster function. Then, you'll use zonal statistics to calculate areas of eligible

open space at a parcel level. Finally, you'll create a custom map and table that detail the spatial information that is required for the Community Rating System review process.
https://learn.arcgis.com/en/projects/assess-open-space-to-lower-flood-insurance-cost/

Oso Mudslide: Before and After

In this lesson, you'll take on the role of a county official who is tasked with spreading awareness about a massive landslide that wiped out a rural community near Oso, Washington. Using imagery of the affected area, you'll create a web-mapping application that allows users to easily compare the area before and after the disaster. You also want users to be able to measure the extent of the impact. The app you create should link to additional materials about the landslide.
https://learn.arcgis.com/en/projects/oso-mudslide-before-and-after/

Monitor Real-Time Emergencies

In these lessons, you'll create a real-time operation view using Operations Dashboard for ArcGIS. Your operation view will monitor the locations of police, fire, and medical vehicles throughout the city and the details of incoming 911 calls. It will also show the locations of nearby emergency facilities. First, you'll explore an existing operation view to see how it works. Then, you'll make your own and share it online.
https://learn.arcgis.com/en/projects/monitor-real-time-emergencies/

Find Potential Hurricane Shelters

In this lesson, you'll map schools in the city of Houston, and add web layers that show evacuation zones and Federal Emergency Management Agency (FEMA)-designated 100-year floodplains. Then, you'll remove the schools that are inside these zones, where flooding is more likely. Last, you'll run a suitability analysis and rank each potential site based on several demographic variables.
https://learn.arcgis.com/en/projects/find-potential-hurricane-shelters/

Find Areas at Risk of Flooding in a Cloudburst

In this lesson, you'll examine and run a geoprocessing model that uses elevation data to find the locations of "bluespots" (low-lying areas that have no natural drainage). For more experienced users, a second model also calculates the volumes of bluespots and their capacity to hold sudden influxes of water. The models are independent: you can do either one or both, depending on your interest and skills. The study area is a municipality near Copenhagen, but the models have global applicability because they are based only on the terrain surface and building location.
https://learn.arcgis.com/en/projects/find-areas-at-risk-of-flooding-in-a-cloudburst/

Predict Floods with Unit Hydrographs

In this lesson, you'll create unit hydrographs for an outlet on the downstream end of the Little River. First, you'll prepare elevation data, and then use it to determine the watershed area for the outlet. Based on your watershed and terrain data, you'll create a velocity field,

which determines how fast water tends to move in your study area. Using the velocity field, you'll create an isochrone map, which assesses the time it takes for water to travel to the outlet from anywhere in the watershed. Last, you'll use the isochrone map to derive a unit hydrograph and interpret what it says about the potential for floods in Stowe, Vermont.
https://learn.arcgis.com/en/projects/predict-floods-with-unit-hydrographs/

Investigate Emergency Response Times
This tutorial presents an approach to using ArcGIS to find out which areas are within a four-minute drive time of a fire station in a city. Although the data is real, the scenario, analysis, and resulting decisions are hypothetical. The purpose of the example is to illustrate the type of problem that can be addressed using drive-time areas.
https://desktop.arcgis.com/en/analytics/case-studies/which-areas-are-within-four-minutes-of-a-fire-station.htm

Maximize Fire Protection Coverage
This tutorial presents an approach to using ArcGIS to find the best location for a new facility (a fire station, in this case). The process uses location-allocation analysis to identify the site that maximizes fire protection coverage. Although the data is real, the scenario, analysis, and resulting decisions are hypothetical. This is just one approach to using spatial analysis tools for siting a new facility. Your criteria and decision-making process might differ.
https://desktop.arcgis.com/en/analytics/case-studies/which-site-for-the-new-station-maximizes-fire-protection-coverage.htm

Chapter 4: A New Way of Policing

Analyze Crime Using Statistics and the R-ArcGIS Bridge
The R-ArcGIS bridge allows you to seamlessly perform analyses on a single dataset by connecting the analytical tools from R and ArcGIS Pro simultaneously. In this lesson, you'll perform statistical analyses to help test the department's theories about the factors that affect unlawful behavior. You also identify the areas where higher numbers of crimes are likely to occur.
https://learn.arcgis.com/en/projects/analyze-crime-using-statistics-and-the-r-arcgis-bridge/

Chapter 5: Making Baltimore Safer

Track Crime Patterns to Aid Law Enforcement
In this lesson, you'll assist the police department of Lincoln, Nebraska, a mid-sized city in the United States. The Lincoln Police Department wants to know how many officers it needs in the field and where they should be concentrated. To do this, first you'll determine the percentage of crime that occurs within a five-minute drive-time area of Lincoln police stations. The crime that occurs outside this drive-time area likely will be handled by field

officers. Then, you'll map the density of crime throughout the city to determine which areas require additional resources and personnel.
https://learn.arcgis.com/en/projects/track-crime-patterns-to-aid-law-enforcement/

Analyzing Violent Crime

This tutorial uses real data to demonstrate the utility of a number of spatial analysis methods, including hot-spot analysis, feature overlay, data enrichment, space-time pattern mining, attribute query, and spatial selection.
https://desktop.arcgis.com/en/analytics/case-studies/broken-bottles-1-overview.htm

Prepare and Present Crime Statistics for a CompStat Meeting

In this lesson, you'll deploy the ArcGIS Crime Analysis solution to analyze robberies in preparation for a CompStat meeting for the city of Naperville, Illinois. After you create the analysis output, you'll publish the results as web layers, and then add them to web maps to create a story map. Your story map will be an Esri Story Map Series that consists of several tabs configured to showcase analysis results and findings.
https://learn.arcgis.com/en/projects/prepare-and-present-crime-statistics-for-a-compstat-meeting/

Chapter 6: CitiStat and the Enterprise of Governing

Manage Hydrant Inspections

The community of San Diego relies on fire hydrants to help put out fires and keep people and property safe. To ensure that the hydrants work when they are needed, the San Diego Fire Department regularly inspects them. In this lesson, you'll help complete the inspections. Acting as different members from the City of San Diego and the San Diego Fire Department, you'll create and manage hydrant inspection assignments and complete that work in the field. Your goal is to make sure that the fire hydrants around the convention center are all inspected before the end of summer.
https://learn.arcgis.com/en/projects/manage-and-complete-hydrant-inspections/

Oversee Snowplows in Real Time

In this lesson, you'll build a web map in ArcGIS Online that contains real-time data about snowplows in a city in Utah. For the government officials, you'll create an operation view in Operations Dashboard for ArcGIS that combines your map with lists, charts, and other helpful information. For the citizens, you'll create a web app with Web AppBuilder for ArcGIS that clearly and simply communicates the key information about roads and snowplows. While both of your final outputs will contain real-time data, each will be tailored to the specific needs of the intended users.
https://learn.arcgis.com/en/projects/oversee-snowplows-in-real-time/

Chapter 8: Improving Education

Fight Child Poverty with Demographic Analysis
Your objective is to ensure that poverty relief programs are offered where they're most needed. To do so, you'll create a map by adding and enriching a layer of census tract areas with demographic data from ArcGIS Living Atlas of the World. You'll also apply smart mapping and style the layer to show median household income by census tract. Then, you'll change the layer style to show median household income and child population to identify which areas have the most children living in poverty. You'll also configure the pop-ups to make the demographic information easier to read. Lastly, you'll report your findings to top donors and partner organizations by configuring a web app that tells a clear story based on your data.
https://learn.arcgis.com/en/projects/fight-child-poverty-with-demographic-analysis/

Chapter 9: Improving Health and Well-Being

Investigate Prescribed Drugs
In this lesson, you'll assume the role of a Health and Human Services analyst investigating lidocaine prescriptions involving two different doctors—one in Washington State and one in Florida. Your goal, as part of a first look at their prescription histories, is to determine whether enough information exists to justify a formal investigation into one or both doctors for illicit oversubscribing. If your data exploration highlights anomalies, you'll flag the data and share your analysis with a supervisor. First, you'll use Insights℠ for ArcGIS® to construct a workbook so you can examine potential anomalies with lidocaine prescriptions from a single doctor in Washington State. Then, you'll analyze prescriptions from a doctor located in Florida.
http://learn.arcgis.com/en/projects/investigate-prescribed-drugs/

Bridging the Breast Cancer Divide
In this lesson, you'll gain understanding about how mortality rates for breast cancer are higher for black women than for white women in the United States. First, you'll explore maps to see what the mortality rates are for black and white women. Then, you'll map the differences in mortality rates to see where the rates differ. You'll map the ratio of mortality rates to see how much they differ. You'll map significant clusters of higher and lower mortality rate ratios so you can focus on the most problematic areas. Finally, you'll map selected breast cancer risk factors to look for explanations for the clusters.
https://learn.arcgis.com/en/projects/map-breast-cancer-differences-by-ethnicity/

Where Does Health Care Cost the Most?
In this lesson, you will:
- Map Medicare costs by local hospital service regions.
- Explore the spatial patterns of costs to see where they're high and low.
- Perform a hot-spot analysis to see where high and low costs are statistically significant.
Identifying areas of unusually high cost is the first step in understanding what factors

might drive those costs, and then contain them with different approaches to service and treatment.
https://learn.arcgis.com/en/projects/where-does-healthcare-cost-the-most/

Homeless in the Badlands
In this lesson, you'll play the role of a GIS specialist for a North Dakota advocacy group. Your group wants to alert state and federal officials to homelessness, a negative effect of the oil boom, to receive more resources for housing and shelters. You'll first download tabular (table-based) data on homelessness. Then, you'll map the data in ArcMap™ by joining it to state geography. You'll publish your map layer to ArcGIS Online and make a set of online maps anyone can access. Finally, you'll present your maps in a single, easy-to-understand web app.
https://learn.arcgis.com/en/projects/homeless-in-the-badlands/

Analyze Traffic Accidents in Space and Time
The techniques demonstrated in this tutorial include tools to detect space-time patterns and methods for both measuring and visualizing 3D hot spots. You will also learn about creating charts to visualize your data and about building models to help you automate your work.
https://desktop.arcgis.com/en/analytics/case-studies/analyzing-crashes-1-overview.htm

Chapter 10: Restoring Our Waters

Model Water Quality Using Interpolation
In this lesson, you'll learn how to analyze water quality data collected in estuaries. Using point measurements collected from across the Chesapeake Bay, you'll create maps of dissolved oxygen levels for the entire Bay for the summer months of 2014 and 2015. First, you'll explore the data using interactive charts to learn the general patterns and characteristics of the data. Then, you'll interpolate these points into smooth maps showing dissolved oxygen levels. Interpolation is the process of taking data measured at specific locations and predicting the value of the dataset at every location between the measurements. Next, you'll perform interpolation using the Geostatistical Wizard, a guided step-by-step environment for performing spatial interpolation, and the Kernel Interpolation With Barriers tool. You'll then learn how to assess the quality of an interpolation model and how to compare two models. After completing the comparison, you'll share your findings with colleagues by making a layout poster showing your results. This poster would be ideal for use at conferences and other gatherings of experts.
https://learn.arcgis.com/en/projects/model-water-quality-using-interpolation/

Connect Streams for Salmon Migration
In this lesson, you'll examine dams along the Mersey River Watershed to determine whether there are any locations suitable for the construction of a fishway. After you identify the best potential location, you'll calculate the dam's upstream watershed to help

determine how much additional habitat could be made accessible by constructing the fishway.
https://learn.arcgis.com/en/projects/connect-streams-for-salmon-migration/

No Dumping: Drains to Ocean
In this lesson, you'll learn how to find the area that drains to a storm drain and the route that pollutants will take if they are dumped or washed into the drain. You'll find the upstream drainage area, called a watershed, for a storm drain near Blackman Elementary School in Tennessee. Then, you'll find the downstream flow path to where it empties into the Gulf of Mexico. Knowing how to find the watershed and flow path, you can experiment to find the watersheds and flow paths from other storm drain locations.
https://learn.arcgis.com/en/projects/no-dumping-drains-to-ocean/

Predict Seagrass Habitats with Machine Learning
You are a marine ecologist who wants to model suitable locations for seagrass habitats around the world. Though you only have seagrass data for a small region of Florida, luckily seagrasses tend to grow in similar ocean conditions in coastal areas around the world. Using the predictive powers of a machine learning model along with the spatial analysis capabilities of ArcGIS Pro, you'll find suitable locations for seagrass growth globally. First, you'll create a training dataset with all the ocean variables that influence seagrass growth. Then, you'll put the variables into Python and use a random forest prediction model to determine where the ocean conditions support seagrass growth. Finally, you'll save the prediction results as a feature class and import it into ArcGIS Pro to find where the highest density of growth is likely to occur.
https://learn.arcgis.com/en/projects/predict-seagrass-habitats-with-machine-learning/

Chapter 11: Preserving Our Land

Deploy an ArcGIS Solution to Build a Park Locator
The City of Naperville, Illinois, has many beautiful parks. The Parks and Recreation Department wants the public to have better access to them. To help their citizens locate the parks nearest to them, they have found an ArcGIS® Solutions template that will allow them to load their data into a pre-configured app with minimal work. In this lesson, you'll download the Solutions add-in, a tool for ArcGIS Pro that allows you to access more than 450 ready-made solutions. This add-in only needs to be downloaded once; after installation, you can access any solution from ArcGIS® Desktop. Then, you'll deploy the Park Locator solution and configure it to match your data. Finally, you'll add your data to the web layers you downloaded with the solution and configure your app.
http://learn.arcgis.com/en/projects/deploy-an-arcgis-solution-to-build-a-park-locator/

Model Soil Contamination
In this tutorial, you will explore data on cesium-137 soil contamination from 1992 in an area near the site of the Chernobyl Nuclear Power Plant to find which kriging model is the

most appropriate for predicting cesium-137 across the region. Kriging is the best available interpolator model that additionally gives prediction uncertainty.
https://desktop.arcgis.com/en/analytics/case-studies/an-appropriate-kriging-model-for-soil-contamination.htm

Classify Land Cover to Measure Shrinking Lakes

In this lesson, you'll assume the role of a geospatial scientist tasked with calculating the change in area of Lake Poyang between 1984 and 2014. Using Landsat imagery, you'll classify land cover in three images of the lake taken at various times over thirty years to show only the surface area of the lake. You'll then determine the change in lake area over time.
https://learn.arcgis.com/en/projects/classify-land-cover-to-measure-shrinking-lakes/

Build a Model to Connect Mountain Lion Habitat

In this lesson, you'll identify potential wildlife corridors that avoid freeways and follow prime mountain lion terrain in order to connect the fragmented mountain lion populations and habitats. First, you'll download a project package with data about the Los Angeles area and open it in ArcGIS Pro. After identifying the criteria that define optimal mountain lion habitats, you'll prepare the data for analysis. Then, you'll create two geoprocessing models: one to determine which areas are suitable habitats for mountain lions, and one that uses the output from the suitability model to generate corridor lines between known mountain lion population centers. By identifying the most suitable locations for wildlife corridors, the cougar population can increase its range and breeding opportunities, improving genetic diversity and survival of the species in Southern California.
https://learn.arcgis.com/en/projects/build-a-model-to-connect-mountain-lion-habitat/

Mapping the Public Garden

In this lesson, you'll first learn how to use the basic mapping tools in ArcGIS to build a pleasing basemap of the garden layer by layer. You'll then publish the basemap to the web as a tiled map service, using custom scale settings to allow extreme close-up zooming. Finally, you'll publish the individual plot boundaries as a feature service that can be accessed on a plot-by-plot basis in the finished map product—a mobile-friendly email application.
https://learn.arcgis.com/en/projects/mapping-the-public-garden/

Chapter 12: Protecting Our Air

Perform a Site Suitability Analysis for a New Wind Farm

A Colorado-based wind energy company has hired you to identify several new sites within the state for the installation of high-efficiency wind turbines. Selecting an appropriate site is key to the success of any renewable energy project, playing a crucial role in financial returns and ease of construction, as well as ongoing operations, maintenance, and overall safety. The American Wind Energy Association (AWEA) has identified these six criteria as crucial in the siting of wind energy projects: adequate wind, land rights, permits, access to transmission lines, a buyer for the wind power, and financing.

http://learn.arcgis.com/en/projects/perform-a-site-suitability-analysis-for-a-new-wind-farm/

Acknowledgments

Whatever good I have been able to do in public service is attributable to the support of my family; the goodness of the people we served; and the commitment, talent, and hard work of the people who chose to serve alongside me.

I want to first and foremost acknowledge the love, support, and sacrifice of my wife, **Katie**, who has affirmed and encouraged my passion for public service from our first date to now. I also want to acknowledge and thank our children, **Grace**, **Tara**, **William**, and **Jack**. No couple in public life ever had more loving or courageous kids.

To the people of Baltimore and the people of Maryland, I will always owe a tremendous debt of gratitude for the honor of serving as their mayor and then as their governor—through often-difficult and challenging days.

I especially want to thank all those men and women who served on the command staff of the successive city and state administrations I had the honor to assemble and lead.

A special thanks to those who served so diligently on my personal staff: **Rebecca Mules**, **Charline Gilbert**, **Michelle Waller**, **Connie Barnes**, and **Linda Aldredge**. They kept a kinetic schedule going every single day with calm, professionalism, and responsiveness to all. To **Connie Barnes**, **Christina Caldwell**, **Julie O'Brien Scheide**, and **Karen Taylor**, whose professional attention to details, follow-up, and duty kept us going. And to all my colleagues and the staff at the Baltimore City Council, who taught me so much as a young council member, including my 3rd District staff: **Tanya Russell**, **Kathy Greenberg**, **Mary Adams**, and **Pamela Carter**.

To **Michael Enright**—my best friend since high school, best man at my wedding, and the best first deputy mayor and chief of staff any mayor-turned-governor could have. To **Peggy Watson**, who returned to public service to become the most needed and effective city finance director in modern Baltimore City history. And to her deputies **Ed Gallagher** and **Helene Grady**, who never saw a budget they couldn't balance.

To **Matt Gallagher**, the very first CitiStat director of any city in the world, and whose passion, discipline, and relentlessness lit our way through even the hardest times. To that dedicated first group of pioneers in a new way of governing—Baltimore deputy dayors **Jeanne Hitchcock**, **Laurie Schwartz**, and **David Scott**; chief of staff **Clarence Bishop**; and city solicitor **Thurman Zollicoffer**— for their courage in returning again and again to the cause of making our city a better place. To all the department heads and bureau chiefs of Baltimore City who hung in there as we quickly moved from routine accountability, to more effective collaborations, to nation-leading results for our people.

I also want to thank a few key elected leaders who helped me when they didn't have to: Former mayor **Tommy D'Alesandro**, for his mentoring and counseling. And for stepping up with Reverend **A.C.D. Vaughn** to save our Housing Authority from receivership and to improve living conditions for all our residents. Former governor **Parris Glendening** and lieutenant governor **Kathleen Kennedy Townsend** for providing the huge increase in funding necessary to free so many of our neighbors from the chains of drug addiction. Former mayor **Kurt L. Schmoke** for the kindness and graciousness that he and Dr. **Patricia Schmoke** showed to me and Katie in transition, and for his friendship, support, and encouragement in the years since. Thank you to house speaker **Mike Busch** and senate president **Mike Miller**. And to **Nancy Pelosi** and **Barbara Mikulski** for never forgetting Baltimore. And to **Gary Hart**, for teaching me that one person can make a difference, and that each of us must try.

I want to thank **Steve Kearney** for his heart and pen in preparing remarks for all occasions for so many years, as well as for his expertise in guiding the communications functions of our city and state governments. And to **Steve Rabin**, speechwriter extraordinaire and his able assistant, **Jackie Small**, along with **Sean Enright** and **Nick Stewart**, for the poetry of heavily footnoted State of the State addresses. To **Paul Hawken** for his patient transcontinental redrafts, which kept spurring us on to better. And thanks to **Frank Perrelli**, whose graphic artistry not only won awards, but also made it easier for citizens to access critical government services.

To **Colm O'Comartun**, with whom I probably spent more waking hours than any other person during the better part of ten years. In one human being, Colm oversaw schedule, advance, and a few different departments. Colm was also a walking encyclopedia of Maryland, US, and international politics. He was, essentially and indispensably, the secretary to the cabinet who traveled with me. It was a great model for us. But I don't think another individual could do all the things that Colm did so very well. And to **Chris Reith** and **Collin Wojciechowski**, for their dedication to duty.

To the leaders of Baltimore's business community who refused to accept the notion that Baltimore wasn't capable of tackling big challenges. In particular, to **Mike Cryor** and **Wally Pinkard**, who stepped up to lead our Believe campaign. And to **John Morton**, **Arnold Williams**, and **Jim Shea**, whose passion and concern for the lives of our poorest citizens inspired other business leaders to support and join the fight.

Thank you, posthumously, to **Jack Maple** for so many things—most importantly for agreeing to come to Baltimore to save lives even as his own was slipping away. To his business and communications partner, **John Linder**, for hearing the goodness within Baltimore and thereby designing the Believe campaign. And for his coaching in the development and sanding of my very rough executive leadership skills. And to my loyal and true friend, **Sue Casey**, who kindly came to town from Colorado at critical times to sort us out and remind us that we need to talk with each other if we are to achieve big things.

I'm especially grateful to **Ralph Tyler**, **Liz Harris**, and their teams of talented and committed public lawyers (**Christina Wellons**, **Eric Beane**, **Meghan**, **Jon Kucskar**, **Josh Auerbach**, and **Mary Pollack**) who safe-guarded—always—the letter and spirit of the law in our pursuit of the greater good.

To Baltimore's first chief of information technology, **Elliot Schlanger**, who put CitiStat together with a 311 call center —an innovation in government that became a model for cities across our nation and around the world.

To **Sean Malone**, in his many roles, but none tougher or more important than sorting out—for a time—the broken disciplinary systems of the Baltimore Police Department.

To **Kevin Enright** and **Peter O'Malley**, who saw that their older brothers were in a tough fight and jumped right in. And to **Brian Hammock**, who rose from loyal intern to talented leader in our years together.

With gratitude and admiration, I salute the leadership and heart of those outstanding leaders who all succeeded **Matt Gallagher** in taking over the reins of CitiStat and then StateStat over successive stretches of time. To **Christopher Thomaskutty** who kept CitiStat going, and who successfully transitioned it into our successor, **Sheila Dixon's**, administration. To **Beth Blauer**, who, like Jack Maple, was rightly and greatly promoted—from juvenile probation officer, to become the longest-serving director of StateStat. And to **Matt Power**, who took over from Beth with the energy we needed to close strong. I'm grateful to **Jay Sakai**, who at the outset of CitiStat quickly turned the Department of Public Works' GIS system into our first CitiStat basemap simply by walking his own GIS know-how across the street. (Jay would later walk back across Fayette Street to become the chief of the Bureau of Water and Wastewater.)

Big shout-out of thanks to the rest of the original **CitiStat team**: **Sam Snowden**, **Andrew Lauland** (who later became director of homeland security), **Kim Amprey** (who would go on to serve as director of Recreation and Parks), **Marla Johnson, Steve Sharkey, Thomas Kim, Catherine Fine, Franklin Branch, Eric Letsinger, Aurpon Bhattacharya, Drew Vetter, Jay Newman, Lindsay Major, Mike Powell** (who would later become chief innovation officer for the State of Maryland), and many others who cycled through for shorter tours of duty, but who went on to do great things all over the country.

And a big thank you to the **StateStat team: Mattie Hutton, Lindsay Major, Matt Raifman, Katherine Klosek, Ronojoy Sen, Sam Sidh, James Bragdon,** and many others who served for only brief times, but with great commitment.

A special thanks to **Izzy Patoka, Angela Bernstein, Tiffany James, Nan Rohrer, Tony Bridges, Elizabeth Weiblen, Gus Augustus, Mark Byrd, Jessy Mejia,** and all the dedicated young men and women of the Mayor's Office of Neighborhoods whose patient listening and many late-night community meetings made our work real in the eyes and lives of our neighbors. And to **Mollie Byron,** our deputy director of Intergovernmental Affairs; to **Iman Awad, Jessy Mejia, David Sloan, Chris Uhl, Jonathan Jays-Green, Yi Shen, Chuck Cook;** and all those who worked in the Governor's Office of Community Initiatives, my heartfelt thanks. One Maryland—where our diversity is our greatest strength. And to all the totally responsive and caring people who worked in the Governor's Office of Constituent Services, led by **Jeremy Rosendale,** thank you.

Others whose talents and commitment were outstanding include **Doug Austin, Richard Burton, Michael Braverman, Denise Duval, Peggy Daidakis, Aaron Greenfield, Eric Letsinger, Bob Maloney, Bill Pencek, Otis Rolley, Reggie Scribber,** and **Keith Scroggins.**

A heartfelt thanks to **Brian Morris, Sam Lloyd,** and **Luwanda Jenkins** whose leadership allowed us to deliver the highest minority and women business inclusion goals in the nation.

I am very grateful to **John Griffin,** who—in addition to leading the Bay Cabinet and BayStat as secretary of the Maryland Department of Natural Resources—also stepped up to serve as my final chief of staff as governor. Deep gratitude to **Steve Neuman** for coming aboard as my senior advisor to secure every gain we could in the final years of my second term. To Deputy Chiefs of Staff **Ashley Valis, Cassie Motz,** and **Ted Dallas,** who made sure that life-saving and life-giving collaborations became the new norm in our state government.

I also wish to acknowledge all the men and women of the ever-diverse and evolving cabinets of those fifteen years of city and state administrations—they were each a vital part of that collaborative circle of leaders who solved problems every day and got things done for the people. I salute also the hardworking men and women of Baltimore city government and the State of Maryland—especially our first responders and National Guard who willingly put their lives on the line for the rest of us every day.

A few outstanding city cabinet leaders are particularly deserving of mention. To **Kristen Mahoney,** who was there from the start at the BPD and concluded her state service as the best criminal justice coordinator any state ever had. To **George Winfield,** who brought integrity, heart, and that Baltimore ethic of hard work to the leadership of the Department of Public Works—I never served with a more decent and kind gentleman. To Housing Commissioner **Paul Graziano,** for his expertise and tireless devotion to the colossal challenge of turning around Baltimore's Housing Authority and Community Development efforts. To **Jay Brodie,** for his unrelenting optimism and belief in what a great city could be. Thank you, **Karen Sitnick,** for all the people for whom you found a job. Thank you, **Joe Kolodziejski,** for making Baltimore cleaner. Thank you, **Bill Gilmore,** for promoting the life, music, and art of Baltimore. Thank you to **Carla Hayden,** the dynamic leader of our Enoch Pratt public libraries who always believed in Baltimore and now runs the Library of Congress.

I'm grateful to **Anthony Brown** for agreeing to run with me for lieutenant governor, and for his outstanding leadership on the issues of job creation, health care, and affordable college over our eight years of service together.

I want to acknowledge some exceptional leaders among the state cabinet secretaries who served the people of Maryland in the O'Malley/Brown Administration: **Eloise Foster**, whose financial expertise brought us through the recession with a triple-A bond rating and a faster rate of job creation than the neighboring states of our region. **Gary Maynard**, who took over one of the worst correctional systems in the country and made it safer and better while also reducing recidivism. **Rich Hall**, our secretary of planning, who greatly advanced the cause of smart growth and the protection of Maryland's rivers, streams, and natural lands. **Marcus Brown**, our superintendent of state police, whose collaborative leadership with local law enforcement drove crime down to thirty-five-year lows. I would also like to thank **Don Boesch** and **Bill Dennison** of the University of Maryland Center for Environmental Science for calling the scientific balls and strikes every month at BayStat. Also, in the vein of BayStat, I thank Maryland Department of the Environment Secretaries **Shari Wilson** and then **Bob Summers** for all their efforts to protect our land, air, and waters. And to Secretaries of Agriculture **Roger Richardson** and **Buddy Hance**, for working the problem. To **Russ Brinsfield**, for his can-do manner and his perspective at BayStat—farmer, mayor, citizen, and lover of the waters of Eastern Shore. And to former Governor **Harry Hughes** and Attorney General **Joe Curran** for starting and sustaining—well before our time—the long fight to Save the Bay.

And to Secretaries of Transportation **John Porcari** and **Beverly Swaim Staley**, and Maryland Transit Administrator **Paul Wiedefeld**, who understood that mobility is a human right. To General **Jim Adkins** and his leadership of our Maryland National Guard through long, deadly, and protracted engagements in Iraq and Afghanistan—and for his unceasing desire to find the Maryland 400, the regiment who fought so bravely during the battle of Battle of Brooklyn at the outset of the Revolutionary War. And to Department of Labor, Licensing, and Regulation Secretaries **Tom Perez**, **Alex Sanchez**, and the indefatigable **Elizabeth Sachs**; and Commissioners of Financial Regulation **Sarah Bloom Raskin** and her successor **Mark Kaufman**, for their leadership in saving so many homes from foreclosure and so many small Maryland banks from failure.

Special thanks to **Pat McGee** for his leadership at Parole & Probation; and to **Hannah Byron** and **Andre Vernot** at the Department of Business and Economic Development.

For their outstanding leadership in public health, I'd like to thank Dr. **Peter Beilenson**, who sprung from CitiStat to create the collaborative stats of DrugStat and LeadStat; and Mr. **John Colmers**, for his heart and compassion as we transitioned institutionalized clients with developmental disabilities to more independent living and greater dignity. I'd also like to thank Secretary **Cathy Raggio** for her leadership of the Department of Disabilities. And Dr. **Josh Sharfstein**, who, after serving a tour of duty in the city, managed to pull off the near-impossible state and federal negotiations that launched Maryland's successful All-Payer/Pay for Wellness rate-setting initiative.

Tremendous appreciation goes out to the following leaders, as well:

To the state's first chief innovation officer, **Bryan Sivak**, for looking over the horizon to where we were headed. And to his successor in that role and CitiStat alumnus, **Mike Powell**, for bringing greater imagination to our pursuit of important goals.

To **Christian Johansson** and **Dominic Murray**, for reminding us by their leadership every day that there is no good more important to a family than a decent job. And to **David Costello**, for corralling the cats of coordinating agencies into a discipline of delivery for achieving progress toward the big strategic goals.

To **Malcolm Woolf** and **Abby Hopper**, who charted a course to sustainable energy for Maryland.

To **Hatim Jabaji**, now of blessed memory, who pioneered an extensive initiative of energy retrofits and upgrades with imagination and drive—thank you.

To **Tom Hickey** and then **Kevin Large**, thank you for making the Board of Public Works meetings seem not so large after all.

To **Andy Lauland**, with gratitude for his steady, persistent focus on improving our homeland security capacities as a city, and then as a state.

A special thank you also to Secretary of Aging **Gloria Lawlah**, State Superintendent of Schools **Lilian Lowery**, Chancellor of the University of Maryland **Britt Kirwin**, and Secretary of Housing **Ray Skinner**.

To **Joe Bryce**, **Stacy Mayer**, **Lisa Jackson**, **Drew Vetter**, **Jeanne Hitchcock**, **Yolanda Winkler**, and **Rebecca Mules** (again); to **David Stamper**, **Shanetta Paskel**, **Sean Malone**, **Carolyn Quattrocki**, **Kevin Hughes**, **Lamorea Stanton**, **Lisa Eutsler**, **Betty Anderson**, and all of those who worked with the state legislature to pass the most progressive agenda in the nation, from marriage equality to comprehensive gun safety legislation, voting rights, a living wage, and repeal of the death penalty—thank you.

Thank you also to Secretary of State **John McDonough** for always making visiting delegations from around the world feel honored, welcomed, and respected by the people of Maryland and their elected government.

Thank you to **Zoe Pagonis**, who upped our game with social media. And to **Lori Livingston** for taking our social media capacities to next level with StateStat and her assistance with the online precursor of this book—*Letters to the People of Maryland*. Thank you also to **Sam Clark**, not only for his work at CitiStat, but also for his untiring service as my *aide-de-camp*. And to his successor, **Chad Scheller**, who made a quick transition from the United States Navy to state government and life on the highways and byways of Maryland. And to **Chris Reith**, **Kara Turner**, and **Collin Wojciechowski** for their dedication to state and duty.

Thanks to **John Ratliff**, our policy director, who took special care of driving our improvements as a state in education.

Thank you also to **Tom DeWire** and **Kevin Loeb** for their talent and heart. And thanks to **Angela Gibson** for her tireless work with the council.

Special thanks also to **Tony White**, **Rick Binetti**, **Raquel Guillory**, **Rick Abbruzzese**, **Gerry Shields**, **Takirra Winfield**, **Shaun Adamec**, **Christine Hansen**, **Shaina Hernandez**, **Nina Smith** and all the dedicated alumni and alumnae of the O'Malley press operations for the honesty and integrity of all our communications to the public. Tremendous thanks to **Jay Baker**, Baltimore City Hall photographer and then official state photographer, and his handful of fellow photographers, many of whose images appear in this book.

Because we learn by teaching, I especially want to thank Dean **Bob Orr** and Assistant Dean **Nina P. Harris** at the University of Maryland School of Public Policy for allowing me to clarify my own experiences in public service by teaching a course on public administration in the Information Age. To **Mo Elleithee** and the Institute of Politics and Public Service at Georgetown University for the opportunity to be on the receiving rather than the giving end of a Georgetown education for one blessed semester. To Dean **Vincent Rougeau**, Assistant Dean **Mike Cassidy**, and **Lissy Medvedow** at Boston College Law School and the Rappaport Center for Law and Public Service. They gave me the privilege of teaching a law school course on public policy and administration at their great law school—complete with case law and citations.

And finally, I am deeply grateful to all the students of Georgetown University, the University of Maryland, and Boston College, whose compassion, optimism, and fearlessness sustained my own optimism about America's better future over these last couple of years.

Contributors

Contributors: Matt Artz, Matt Ball, Sir Michael Barber, Jim Baumann, Matt Beckwith, Carri Beer, Clint Brown, Hannah Burke, Joan Cramer, Mark Cygan, Jack Dangermond, Nick Dilks, Stephen Goldsmith, Christian Harder, Jim Herries, Michael Hinde, Maria Jordan, Stacy Krieg, Derek Law, Keith Mann, Amen Ra Mashariki, J.D. Merrill, David Nyenhuis, Catherine Ortiz, Matt Power, Monica Pratt, Steve Rabin, Josh Sharfstein, Citabria Stevens, Bern Szukalski, Helen Thompson, David Traversi, Andrew Turner, and Jerry Weast

Editors: Matt Artz, Georgette Beatty, Clint Brown, Sasha Gallardo, Mark Henry, Stacy Krieg, Catherine Ortiz, and Carolyn Schatz

Book design and layout: Matt Artz

Front and back cover design: Monica McGregor

Portions of this book were inspired by my previous writings and presentations, including:
- *Seizing America's Renewable Energy Future.* Sustainability Symposium: Champions of Change, January 8, 2018, Orlando, Florida.
- *A Better Way of Governing: The data, the map, and the method.* Urban Land Institute, October 2017, Dublin, Ireland.
- *Saving the Bay: Actions, not words.* September 2017.
- *Letters from Boston.* Medium, January 30, 2017 to May 8, 2017.
- *The Opportunity of Climate Change: The Fire of the Human Spirit.* Sustainability Symposium: Ready for Anything, January 9, 2017, Orlando, Florida.
- *A New Way of Governing.* TEDxMidAtlantic, October 21, 2016, Washington, D.C.
- Keynote address. Esri International User Conference, July 20, 2015, San Diego, California.
- Keynote address. Esri Federal GIS Conference, February 9, 2015, Washington, D.C.
- *Letters to the People of Maryland.* Tumblr, January 1, 2015 to November 25, 2015.
- *A Prescription for Innovation: Maryland's Data-Driven Approach to Containing Costs and Advancing Health.* State of Maryland, August 2014.
- Address to the Harvard Kennedy School, ARCO Forum, April 19, 2001.

Credits

Front Cover
Photograph by Jay Baker.

Back Cover
Photograph by Jay Baker.
Map by John Nelson.

Introduction
Page 2: Photograph by Jay Baker.
Page 5: Map by Esri.

Chapter 1: A New Way of Governing
Pages 6–7: Photograph by Dawit Tibebu on Pixabay.
Page 10: Illustration by Esri.
Page 11: Illustration by Esri; map by the City of Baltimore.
Page 13: Map by the Office of Unified Communications, District of Columbia.
Page 19: Photograph by Jay Baker.

Chapter 2: When Disaster Strikes
Pages 22–23: Photograph by Ian Froome on Unsplash.
Page 25: Photograph used by permission from Baltimore Sun Media.
Page 26: Photograph used by permission from Baltimore Sun Media.
Page 27: Map by Esri and Frank Perrell, City of Baltimore.
Page 29: Illustration by Esri.
Page 30: Photograph used by permission from Baltimore Sun Media.
Page 31: Photograph used by permission from Baltimore Sun Media.
Page 33: Maps and dashboards courtesy of the State of Maryland, Department of Information Technology.
Page 34: Illustration by Frank Perrell, City of Baltimore; photograph by Bruce Emmerling on Pixabay.
Page 39: Dashboard courtesy of the State of Maryland, Department of Information Technology.
Page 40: Photograph by Jesse Mills on Unsplash.
Page 41: Map by Esri.

Chapter 3: Collaborative Leadership
Pages 42–43: Photograph by Hans-Peter Gauster on Unsplash.
Page 45: Photograph by Greg Pease Photography.
Page 48: Photograph by Jay Baker.
Page 52: Photograph by Jay Baker.
Pages 58–59: David M. Traversi, *The Source of Leadership* (2017).
Page 61: Maps by Esri.

Chapter 4: A New Way of Policing
Pages 62–63: Photograph by NASA on Unsplash.
Page 67: Photograph by Thomas J. Fitzsimmons.

Page 68: Illustration by Esri.

Page 69: Photograph by Meriç Dalı on Unsplash.

Page 70: Graph by Esri; data from the Federal Bureau of Investigation and the City of New York Police Department.

Page 73: Map by Esri.

Page 74: Map by the City of Baltimore.

Pages 76–77: Dashboards by Esri.

Page 79: Map by Esri.

Chapter 5: Making Baltimore Safer

Pages 80–81: Photograph by Bruce Emmerling on Pixabay.

Page 82: Photograph by Scott Beyer/Market Urbanism Report.

Page 84: Graph by Esri; data from the Federal Bureau of Investigation, the City of Los Angeles Police Department, the City of New York Police Department, and the City of Baltimore Police Department.

Page 85: Photograph by Jay Baker.

Page 86: Maps by the City of Baltimore.

Page 88: Map by the City of Baltimore.

Page 91: Maps by the City of Baltimore.

Page 92: Illustration by Esri.

Page 94: Graph by Esri; data from the Federal Bureau of Investigation and the City of Baltimore Police Department.

Page 97: Graph by Esri; data from the City of Baltimore Police Department.

Page 98: Photograph by Jay Baker.

Page 99: Graph by Esri; data from the City of Baltimore Police Department.

Page 100: Photograph by Greg Pease Photography.

Chapter 6: CitiStat and the Enterprise of Governing

Pages 102–3: Photograph by Jay Baker.

Page 105: Photograph by Jay Baker.

Page 106: Photograph by Jay Baker.

Page 109: Photographs by Jay Baker.

Pages 110–11: Illustration by Esri; logo by the City of Baltimore.

Page 113: Graphs by Esri; data from the City of Baltimore.

Page 115: Illustration by Esri; map by the City of Baltimore.

Page 116: Chart by Esri; data from the City of Baltimore.

Pages 120–21: Dashboards by the City of Baton Rouge and Parish of East Baton Rouge, Louisiana.

Page 122: Maps by the City of Baltimore.

Page 123: Photograph by Jay Baker.

Page 124: Map by the City of Baltimore.

Page 125: City of Baltimore.

Chapter 7: Taking It Statewide

Pages 128–29: Photograph by Martin Falbisoner on Wikipedia, CC BY-SA 3.0.

Page 130: Mwaps by Esri.

Page 131: Photograph by Jay Baker.

Pages 134–35: Illustration by Esri; logo courtesy of the State of Maryland, Department of Information Technology.

Pages 140–41: Sir Michael Barber, Delivery Associates.

Pages 142–43: Dashboards courtesy of the State of Maryland, Department of Information Technology.

Page 145: Photograph by Jay Baker.

Page 148: Chart by Esri; data from the State of Maryland.

Page 149: Photograph by Greg Pease/Getty Images.

Page 150: Graph by Esri; data from the State of Maryland.

Page 155: Dashboard courtesy of the State of Maryland, Department of Information Technology.

Chapter 8: Improving Education

Pages 156–57: Photograph by Kimberly Farmer on Unsplash.

Page 158: Photograph by Jay Baker.

Page 160: Graph by Esri; data from the State of Maryland.

Page 163: Data from the State of Maryland.

Page 164: Map courtesy of the State of Maryland, Department of Information Technology.

Page 165: Dashboard courtesy of the State of Maryland, Department of Information Technology.

Page 167: Montgomery County Public Schools.

Page 169: Graph by Esri; data from the State of Maryland.

Page 170: Map developed at Montgomery County Public Schools.

Pages 170–71: Adapted from a chapter titled "Confronting the Achievement Gap: A District-Level Perspective" in *Improving the Odds for America's Children*, a compilation book by Harvard Education Press, 2014; Seven Keys to College Readiness developed at Montgomery County Public Schools.

Page 172: Data from the State of Maryland.

Page 173: Dashboard courtesy of the State of Maryland, Department of Information Technology.

Page 174: Photograph by Jay Baker.

Page 175: Story Map by Esri.

Chapter 9: Improving Health and Well-Being

Pages 176–77: Photograph by Niwat on Rawpixel.

Page 178: Photograph by Greg Pease Photography.

Page 180: Graph by Esri; data by the State of Maryland.

Pages 182–83: Dr. Joshua M. Sharfstein.

Page 185: Map courtesy of the State of Maryland, Department of Information Technology; data by the State of Maryland.

Page 187: Dashboard courtesy of the State of Maryland, Department of Information Technology.

Page 188: Graph by Esri; data by the State of Maryland.

Page 189: Maps courtesy of the State of Maryland, Department of Information Technology.

Page 190: Dashboard courtesy of the State of Maryland, Department of Information Technology.

Page 191: Map by Esri.

Page 192: Chart by Esri; data by the State of Maryland.

Page 193: Illustration by Esri; data by the State of Maryland.

Page 194: Photograph by Greg Pease Photography.

Page 195: Dashboard by the City of Tempe, Arizona.

Chapter 10: Restoring Our Waters

Pages 196–97: Photograph by Dave Harp.

Page 199: Graph by Esri; data by the State of Maryland.

Page 200: Logo courtesy of the State of Maryland, Department of Information Technology.

Page 201: Maps courtesy of the State of Maryland, Department of Information Technology.

Page 202: Photograph by Jay Baker.

Page 203: Maps courtesy of the State of Maryland, Department of Information Technology.

Page 205: Illustration by the University of Maryland Center for Environmental Science.

Page 207: Photograph by Jay Baker.

Page 208: Map courtesy of the State of Maryland, Department of Information Technology.

Page 209: Photograph by Dave Harp.

Page 211: Chris McChesney, Sean Covey, and Jim Huling; illustration by Esri.

Pages 212–13: Maps by the University of Maryland Center for Environmental Science.

Page 214: Map courtesy of the State of Maryland, Department of Information Technology.

Page 215: Chart by Esri; data by the State of Maryland.

Page 216: Photograph by Dave Harp.

Page 218: Dashboard courtesy of the State of Maryland, Department of Information Technology.

Page 219: Map courtesy of the State of Maryland, Department of Information Technology.

Chapter 11: Preserving Our Land

Pages 220–21: Photograph by Greg Pease/Getty Images.

Page 223: Photograph by Jay Baker.

Page 224: Map courtesy of the State of Maryland, Department of Information Technology.

Page 226: This map was produced using data, in whole or in part, provided by The Trust for Public Land.

Page 229: Maps courtesy of the State of Maryland, Department of Information Technology.

Page 230: Nick Dilks, Ecosystem Investment Partners, LLC.

Page 231: Imagery by the USDA Farm Services Agency's National Agriculture Imagery Program.

Page 235: Photograph by Esri.

Page 236: Graph by Esri; data by the State of Maryland.

Page 237: Map courtesy of the State of Maryland, Department of Information Technology.

Chapter 12: Protecting Our Air

Pages 238–39: Photograph by Dave Harp.

Page 241: Photograph in the public domain.

Page 243: Carbon Neutral Cities Alliance.

Page 244: Chart by Esri; data by the State of Maryland.

Page 245: Dashboard courtesy of the State of Maryland, Department of Information Technology.

Page 246: Photograph by Dave Harp.

Page 248: Dashboard courtesy of the State of Maryland, Department of Information Technology.

Page 249: Map by Esri.

Page 252: Michael Hindle and Carri Beer.

Page 253: Dashboard courtesy of the State of Maryland, Department of Information Technology.

Page 254: Graph by Esri; data by the State of Maryland.

Page 255: Map by Esri.

Page 257: Map by BOEM Office of Renewable Energy Programs, MarineCadastre.gov, NOAA.

Chapter 13: The American Revolution, and Our World

Pages 258–59: Photograph by Joshua Earle on Unsplash.

Page 260: Painting by Gilbert Stuart Williamstown/ Crystal Bridges Museum of American Art, Bentonville, Arkansas, 2005.27.

Page 261: Map by George Washington/Library of Congress, Geography and Map Division.

Page 262: Photograph by Department of Housing and Urban Development Staff; US National Archives.

Page 263: Dashboards by Esri.

Page 264: Daniel Kaufmann, Natural Resource Governance Institute (NRGI) and Brookings Institution, and Aart Kraay, World Bank Development research Group.

Page 266: Illustration by the United Nations.

Page 267: Dashboard by the United Nations Department of Economic and Social Affairs.

Page 269: Story Map by the United Nations Statistics Division.

Chapter 14: The Road Ahead

Pages 270–71: Photograph by Pexels on Pixabay.

Page 273: John F. Kennedy Presidential Library and Museum.

Page 275: Illustration by Esri.

Page 276: Graph by Esri; data by the State of Maryland.

Page 277: Illustration by Esri and Sean W. Malinowski, PhD.

Page 279: Illustration by Esri and Grant Thornton, LLP.

Page 281: Maps by Eagle Technology.

Page 282: Illustration by Esri.

Page 286: Photograph by Bruce Emmerling on Pixabay.

Page 287: Map by the City of Brampton.

About the Author

Just two years after Martin O'Malley's upset election as mayor of Baltimore in 1999, *Time* magazine called him one of America's best big-city mayors. His new data-driven system of performance management, CitiStat, earned his city the Innovations in Government Award from the Harvard Kennedy School, and has been copied by mayors across the country and around the world. When he ran for his party's nomination for president in 2016—after two successful terms as governor of Maryland—*Washingtonian* magazine called him "the best manager in government today."

O'Malley was the first of a new generation of "smart city" mayors who would follow. CitiStat—and its Maryland progeny, StateStat—inspired key amendments to the Government Performance and Results Act, foundational requirements intended to spread data-driven management practices across federal agencies today.

In *Smarter Government*, O'Malley lays out how to govern for better results in the Information Age. Every elected leader can use this formula to make government services more efficient and effective. But it requires a commitment to transparency, the courage to follow the data wherever it might lead, and the resolve to measure the way government works. It is all about producing better results—in real time, for real people.

O'Malley and his wife, Katie, live in Baltimore. He has lectured on public administration and the rise of smart cities at the University of Maryland, the Georgetown University Institute of Politics and Public Service, Boston College School of Law, and the Harvard Kennedy School. He chairs the advisory board of the MetroLab Network, a nonprofit of fifty leading US cities and their university partners. He also advises a number of technology companies in the areas of government technology, transportation, and communications, and serves as a smart-cities adviser to Grant Thornton LLP.

$\int$